FORMAL ASPECTS OF VLSI DESIGN

FORMAL ASPECTS OF VLSI DESIGN

Proceedings of the 1985 Edinburgh Workshop on VLSI
Edinburgh, Scotland, U.K.

editors

George J. MILNE
University of Edinburgh
Edinburgh, U.K.

and

P.A. SUBRAHMANYAM
AT & T Bell Laboratories
Holmdell, NJ, U.S.A.

1986

NORTH-HOLLAND
AMSTERDAM · NEW YORK · OXFORD · TOKYO

ISBN: 0 444 70026 9

Published by:
ELSEVIER SCIENCE PUBLISHERS B.V.
P.O. Box 1991
1000 BZ Amsterdam
The Netherlands

Sole distributors for the U.S.A. and Canada:
ELSEVIER SCIENCE PUBLISHING COMPANY, INC.
52 Vanderbilt Avenue
New York, N.Y. 10017
U.S.A.

Library of Congress Cataloging-in-Publication Data

Edinburgh Workshop on VLSI (1985)
Formal aspects of VLSI design.

1. Integrated circuits--Very large scale integration--Design and construction--Congresses. I. Milne, George J., 1952- . II. Subramanyam, P. A. III. Title.
TK7874.E28 1985 621.395 86-8836
ISBN 0-444-70026-9

PRINTED IN THE NETHERLANDS

FORMAL ASPECTS OF VLSI DESIGN

Proceedings of the 1985 Edinburgh Workshop on VLSI

The collection of papers in this book is representative of the discussions and presentations at the 1985 Workshop on Formal Aspects of VLSI Design held at Edinburgh, and embodies ongoing work at universities and industrial research laboratories, both in Europe and North America. The research described in this proceedings recognises the increasing relevance of formal techniques for synthesis, verification, and simulation in the context of VLSI design. While the motivations for bringing formalism to bear upon these problems are many and varied, the major rationale is the need to cope with the complexity of the VLSI design problem while guaranteeing a degree of rigour to the design process such that correct designs evolve.

The papers presented here are representative of formalisms being used in VLSI for a fairly broad spectrum of activities – design, simulation, verification, and layout. However, they may be broadly classified into efforts directed at (1) provably correct design proceeding from formal specifications, and (2) a posteriori verification (and simulation) techniques aimed at rigorously guaranteeing the correctness of existing designs. The ordering of papers in the book reflects this categorization, and is alphabetical within each category.

Provably Correct Design Evolution

In "Towards Verifiably Correct VLSI Design" George Milne shows how a spectrum of interrelated languages capturing both physical and behavioral properties may be used to describe, design and analyze VLSI circuits; these languages are based on a circuit calculus called CIRCAL. Two verifiable design techniques are discussed: a completely automatic approach which isolates the user from device semantics and verification, and a system in which the designer develops the implementation with the system providing interactive verification. A simple equality circuit is used as an illustration.

In "Design Transformation and Chip Planning", Robert Milne discusses aspects of a prototype system called LTS (for "layout and timing for structures") at Standard Telecommunications Laboratories for supporting VLSI design. A logic-based language is used for specifying structural designs in LTS; the paper elaborates upon the relation between the graphical and textual evolution of a design illustrated in this paper. Milne remarks that introducing formal design in industry might be arduous, since product developers are reluctant to deviate from established and familiar methods and aids.

In "Digital Design in a Functional Calculus", Johnson discusses transformations of recursive function definitions into digital designs; the resulting designs are described at roughly the register transfer level of abstraction. A lazy Lisp-like language called Daisy is used for this purpose. Johnson remarks that his techniques are suited more to the implementation of programming language processors than to the generation of circuits that have either iterative or distributed control, e.g., digital filters or systolic arrays. The generation of circuitry needed to interface, power, and clock the implementation is not addressed.

In "The algebraic basis of an Expert System for VLSI Design", Subrahmanyam outlines an algebraic framework for supporting a broad spectrum of the activities involved in specification, synthesis, and analysis. Transformations are used to map an initial abstract behavioral description of the problem along with (optional) performance requirements into a detailed architectural description at the chip level; such descriptions are then used as inputs to lower level silicon compilation tools for producing fabrication masks. The framework developed provides the ability to reason about the syntax and semantics of computational networks and their interrelationships to performance measures and network geometry; in addition, it supports a gradual introduction of system timing and synchronization information as a design evolves.

A Posteriori Design Verification

In "Specification and VLSI Design" Birtwistle et al. present work in progress on two large proof efforts: one focusses on the elimination unit of a local area network, and the other on an implementation of Landin's SECD machine. The proofs use specifications that are mainly at the register transfer level. Their experience seems to underline the fact that bottom-up are very difficult to perform, and reinforces the importance of top-down specification driven design methodologies.

Boute's paper describes work on second-order system semantics for digital systems. The functional programming language SASL is used to illustrate the notions involved. It is argued that the feedback involved in synchronous sequential circuits necessitates second-order concepts be used to maintain useful properties for formal reasoning.

Browne, Clarke and Dill discuss their experience in automatic circuit verification using a model checker for formulas in Computation Tree Logic. Two examples of proofs are outlined: an asynchronous queue element, and the control circuitry for an asynchronous communications interface adapter similar to the 6850. The authors argue that the ability of the model checker to produce a counterexample in the cases when a given formula is unsatisfiable is useful in debugging implementations.

In "Can a Simulator Verify a Circuit?", Bryant examines conditions under which a logic simulator can be used to prove the correctness of a circuit, and explores methods for verifying circuits using a three-valued (0,1, and X or undetermined) logic simulator.

Eveking, in "Formal Verification of Synchronous Systems" discusses the use of predicate calculus to show that an implementation consisting of a synchronous circuit using gates and flipflops is consistent with a specification expressed as a sequential machine.

Gordon cogently argues that higher-order logic is a good formalism for specifying and verifying hardware. The specification and verifications techniques are illustrated via a series of examples. The underlying proof system used is HOL, a machine oriented formulations of higher-order logic.

Hanna and Daeche use higher-order logic as a hardware description language both for rigorous behavioral specification and for the formal verification of digital device correctness. Their VERITAS system permits specification and verification based on the use of higher-order logic and it is presented using an edge-triggered D-type flipflop example.

We believe that this is one of the first attempts to bring together work directed at applying formal verification and synthesis techniques to the area of VLSI design. It has been our experience that such attempts result in a cross fertilization of ideas that yield new theoretical insights; this fact is an added incentive for conducting research in this area. As should be obvious from an examination of the papers, there is no dearth of challenging problems in this area. It is our sincere hope that this book will provide stimulus to this relatively new but important field.

George Milne
P.A. Subrahmanyam

The 1985 Edinburgh VLSI Workshop was made possible by sponsorship from: The Scottish Development Agency, The Alvey Directorate, EFCIS, GEC, IBM, Lattice Logic, National Semiconductor, Plessey Research, Racal Electronics, Systems Designers International, Texas Instruments. The 1985 Workshop was the fourth in a series held annually at the Computer Science Department, Edinburgh University; past workshop topics have included Silicon Compilation, CMOS for VLSI and Expert Systems for Design.

TABLE OF CONTENTS

Formal Aspects of VLSI Design
G.J. Milne and P.A. Subrahmanyam (editors)

Towards Verifiably Correct VLSI Design

George J Milne
Computer Science Department
University of Edinburgh
Edinburgh, U.K.

The incorporation of behavioural verification into the design process is identified as a new approach to VLSI systems design. It is shown how a description language spectrum capturing both structural and behavioural properties may be used to describe, design and analyse complex VLSI circuits. Two verifiable design techniques are illustrated; a purely syntactic approach involving automatic design generation which isolates the user from device semantics and verification, and a system in which the designer is required to describe both behavioural and structural features with the system providing interactive verification as the design evolves.

1. INTRODUCTION

The validation of device behaviour before fabrication is crucial to the complete VLSI design process. Simulation techniques are commonly used to "execute" some form of design representation, simulating the behaviour of the intended chip. This simulation is then examined to see whether the given design meets its specification, allowing a proportion of design faults to be detected and remedied before fabrication. Simulation is an unsatisfactory technique for validating designs, since like testing it can be used to show the presence of errors but can never conclusively demonstrate their complete absence. An alternative approach to design validation is presented; that of verifying design correctness within a model of device behaviour by mathematical proof rather than by informal reasoning.

The CIRCAL model [4],[6],[7] is used both to represent device behaviour and to support the proof techniques [8] which allow us to demonstrate that a constructed device satisfies its specified behaviour. CIRCAL also provides a basis to a VLSI design language involving a spectrum of representation levels which are passed through as a design evolves hierarchically. Rather than attempt to verify the correctness of a complete chip by a single proof we take advantage of the design hierarchy to integrate verification with design. This step-wire process involves a series of proofs which are performed as the hierarchical design evolves.

This paper presents two techniques for verifiable design:

- A "silicon compiler" may be used to automatically produce correct designs from a source description. This approach buffers the designer from low-level design issues, from the need to specify the intended behaviour in detail and from verification issues. The automated design process will itself be proved correct with only correct designs being produced.

- Alternatively, a skilled designer will wish to be more intimately involved in the design process by specifying both spatial (architecture and layout) and behavioural (function, logic and timing) properties of the design. An interactive verification system will establish design correctness as the design evolves, and this technique is presented in detail by means of an example.

2. BACKGROUND

VLSI design tools have evolved from an approach which relies on interpreting an informal description of the intended device as a circuit diagram and then transforming it directly into a physical layout. Validation of the design to ensure that it correctly represents the original requirement is achieved by testing the fabricated chip itself. The increasing complexity of integrated circuits renders this simple approach impractical; the steps between specification, circuit diagram and layout are wide and they expose the designer to error generation which can only be detected and corrected by testing the chip and performing a redesign and refabrication if necessary.

Current design techniques have introduced intermediate design representations such as logic diagrams, floor plans and stick diagrams to narrow the size of the design steps. This increases confidence that each step has not introduced design decisions which fail to satisfy the original specification. Features such as graphics, layout languages in the style of Mead and Conway [3], design rule checking algorithms, PLA generators and gate array and standard cell systems improve our ability to produce correct designs. Together with various formal and informal techniques for analysing circuit timing, simulators are almost universally used as the only practically available technique for validating chip behaviour prior to fabrication. While these features greatly aid the design process they have inherent limitations which become more and more significant as design complexity increases.

Required chip behaviour is usually informally specified causing problems when attempting to (informally) validate a design against the possibly ambiguous specification. The intermediate design representations tend to be too low-level with the result that design steps are still too large. The various intermediate design representations available are unrelated with no process for translating designs in one representation into designs in another. Finally, simulation is unsatisfactory when validating intermediate representations; it can never demonstrate the complete absence of errors while the complexity of VLSI circuits makes exhaustive simulation infeasible.

Integrating design with provable verification in a framework of related design representations is proposed as a major departure from current design systems which, it is hoped, will attack some of the problems mentioned above.

3. A FRAMEWORK FOR DESIGN

To perform mathematically-based verification as an integral part of the hierarchical design process requires the formulation of a spectrum of related description languages. These will range from a high-level language describing the function of the intended device to a mask-level, geometrical language from which fabrication can proceed. Each representation language, or level of description, will be used to describe a design at a given level of detail and various design paths, or mappings, will exist between the levels.

A spectrum of description languages and the mappings between them is shown in Figure 1.

It is intended that a design system incorporating this description framework will be used as follows.- A designer works down the spectrum from a formal description of the functional behaviour of the intended chip. To add more information the design passes to a lower level with design decisions being made either by the designer, if skilled enough, or automatically. Each level down captures one more feature required in the design and this ensures that the mapping between levels

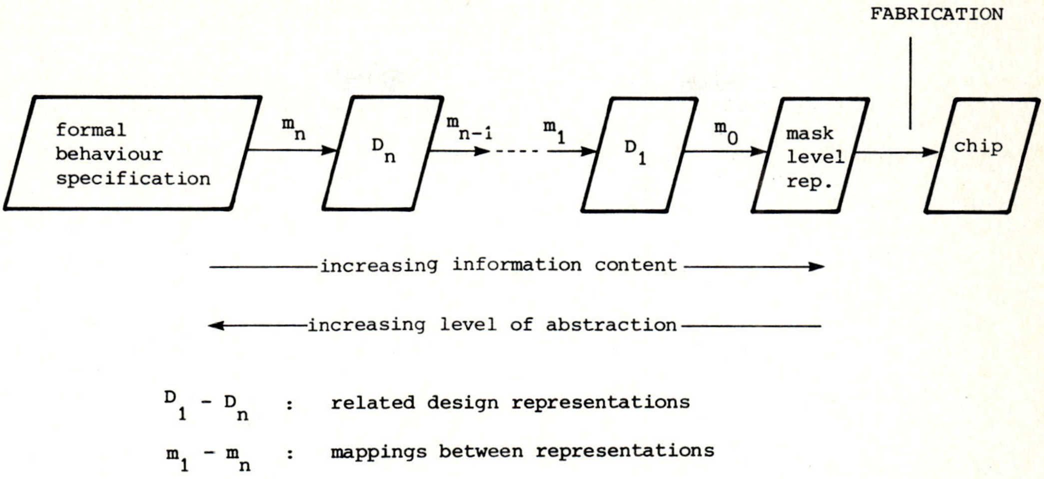

Figure 1

are kept simple, gives structure to the design process and facilitates proof of each design step. Alternative paths through the design levels could also be taken, reaching a low-level representation by linking through a different ordering of levels; designing one feature before the other depending on the design exercise being performed.

The skilled designer will craft his design by passing down the spectrum manually. At the lower levels this exploits the available technology to the full to achieve an efficient design with respect to properties such as time and area. A less skilled designer will wish to make full use of automated design algorithms and a sequence of such design steps can be coalesced into a single automatic procedure. This is the "silicon compiler" approach whose ultimate objective is to go from a high-level representation directly to chip fabrication without human intervention.

A top-down design process repeatedly takes a design at some level of description and implements it in terms of simpler component parts with decisions on architecture, topology and function being repeatedly made. Within each level heterarchic design techniques are employed with components described at a similar level of abstraction being placed and suitably connected. Only when we have designed components which cannot be further "simplified" at the current level do we progress to the next lower level. As we pass down the design spectrum we will reach a level where the required components already exist in a database of basic building blocks and the top-down design has then met with the bottom-up layout process.

Each design step would be validated to ensure that the resulting implementation satisfies the original specification. Formal verification requires a proof that every successive design description is connected by a correct mapping; that is a mapping which preserves relevant behavioural properties. If every step in the design spectrum can be verified then a proof that the resulting chip is correct with respect to the highest level specification has been performed. Failure to verify a design step correct forces iteration; repetition of preceding design steps using different design choices.

4. VERIFYING DESIGN CORRECTNESS

A mathematical framework, called CIRCAL [4],[6],[7] is used to denote the behaviour of a design expressed in one of the levels in the design spectrum. Each level has its semantics unambiguously defined by a function which maps from its syntax into the CIRCAL model. The meaning of a particular design is given by a CIRCAL expression which is formed algebraically; the meaning of a design constructed out of a number of parts is given by a suitable composition of the CIRCAL expressions which denote the component parts.

It is the ability to formally express the meaning of designs constructed in the various description languages which allows us to verify individual designs and the design process itself. To prove the correctness of a mapping between designs expressed in adjacent levels of representation involves using two semantic functions to give meaning to the higher level design and to the design which results from applying the mapping to it, respectively. The two resulting CIRCAL expressions denote behavioural details at the adjacent levels. The correctness of the mapping is then established by proving that the two terms are equivalent after performing certain abstractions on the lower-level expression which balance up the information content of both expressions.

The verification of a single step in the design process is illustrated in Figure 2. A specific design d created at description level D_i is mapped into level D_j using design step m_j; design $m_j(d)$ results and meaning is given to the two designs using functions s_i and s_j respectively.

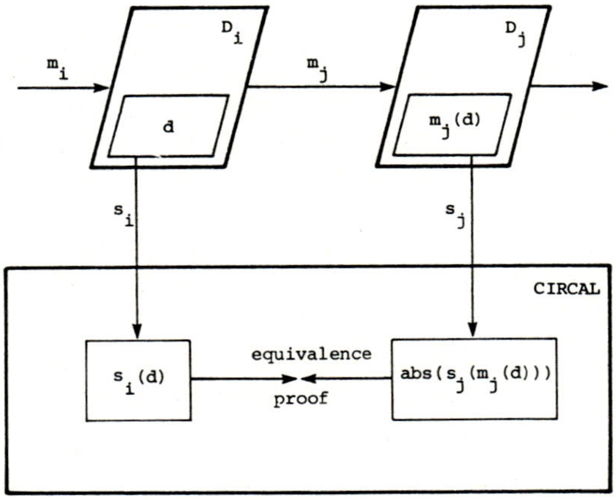

D_i, D_j	:	design descriptions at levels i and j.
m_j	:	mapping from D_i to D_j
s_i, s_j	:	semantic functions giving meaning to designs in D_i and D_j respectively by expressing their behaviour in the CIRCAL model.
$s_i(d)$	:	a CIRCAL expression giving meaning to a specific design in D_i, call it d.
$abs(s_j(m_j(d)))$	:	a CIRCAL expression giving meaning to a specific design in D_j obtained by applying m_j to d.

Figure 2

The equivalence proof

$$s_i(d) \equiv abs(s_j(m_j(d)))$$

establishes that m_j correctly maps design d in D_i into a design in D_j preserving the behaviour at the higher level. An example of this style of verifiable design is presented in Section 6.

If m_j is performed by the designer, with information being added according to how he intends to implement the higher level features at the lower level, then this proof technique would be used every time the design step m_j is made. If m_j automatically produces a design at level D_j from one at level D_i then the possibility exists to prove the compile function m_j correct for all designs which can be created using the syntax of D_i. We would then prove that

$$\forall d \in D_i \,.\, s_i(d) \equiv abs(s_j(m_j(d)))$$

The feasibility of this approach has been demonstrated in [5] where a simple silicon compiler which constructs circuits to implement nor expressions is proved correct. While this example is a simple one, the consequences of this approach should not be underestimated. If CIRCAL is used to verify the correctness of a series of automated mappings across the whole design spectrum, then the possibility exists of establishing the correctness of an ideal, general purpose silicon compiler. Rather than prove every single circuit correct, silicon compiler correctness ensures that the whole family of designs which are automatically produced are correct without requiring the designer to perform the verification task himself, so isolating the user from both design and verification issues.

Figure 3 illustrates how CIRCAL fits into the design spectrum with testing being required to exhibit fabrication errors.

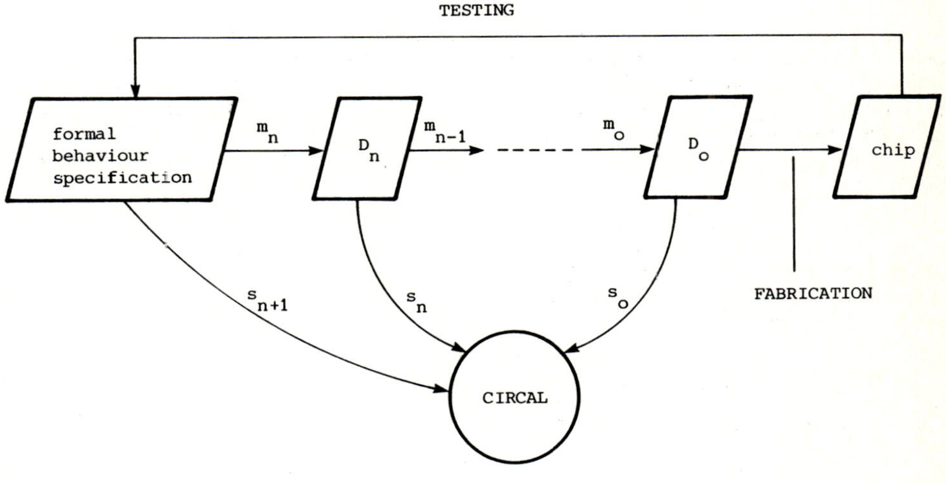

Figure 3

The correctness of a complete silicon compiler translating from a formal behaviour specification to a mask-level representation D_o is established if the diagram in Figure 3 can be shown to commute.

5. THE CIRCAL MODEL

CIRCAL provides a framework in which to represent both the physical structure of a circuit and its behaviour as it evolves through time. The behaviour of a circuit component such as a transistor, a gate or a much larger device is the sequence of events which the component wishes to perform in conjunction with other components in its environment using a number of predetermined communication ports. The linkage conventions of the system being modelled may be shown by picturing devices as boxes with arcs connecting their labelled ports. For a given level of description boxes are in a 1-1 correspondence with the components of the system being modelled.

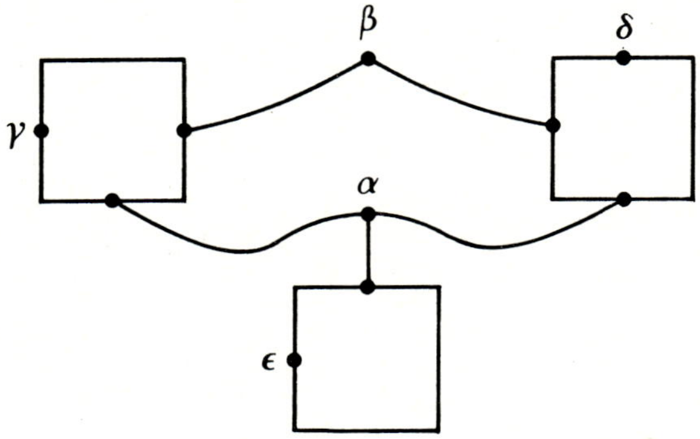

Figure 4

We use the convention that two or more ports with the same label are joined and the labels are moved to the midpoint. The three devices pictured in Figure 4 are all linked via their α port while only two are linked using the β ports. These ports can then be used to attach further devices to construct a larger system.

The three interlinked devices each have their behaviour specified by a CIRCAL expression. Every expression has an associated sort; a set of labels which indicate where a device may interact with others. The corresponding behaviour expressions indicate if, how or when the ports are actually used. The expression denoting the whole system pictured in Figure 4 is obtained by applying the CIRCAL dot operator to the component expressions. Expressions are constructed from operators as listed below.

For D_L being the domain of behaviour expressions with sort L (a set of labels remember) we have the following operators and their type:-

guarding	$\lambda: D_L \rightarrow D_L$ where $\lambda \subseteq L$
choice	$+: D_L \times D_L \rightarrow D_L$
nondeterminism	$\oplus: D_L \times D_L \rightarrow D_L$
termination	$\Delta: D_L$
composition (dot)	$\bullet: D_L \times D_M \rightarrow D_{L\cup M}$
abstraction	$-\alpha: D_L \rightarrow D_{L-\{\alpha\}}$ where $\alpha \in L$

Example: A device of sort $\{\alpha, \beta\}$ is pictured by

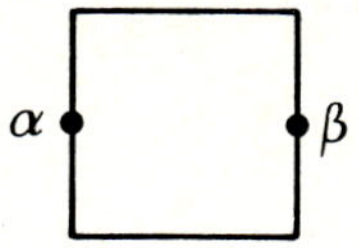

Figure 5

and its behaviour may be described informally by "event α preceeds event β and the device terminates". This behaviour is modelled in CIRCAL by expression

$$\alpha \; \beta \; \Delta$$

Sequentiality is represented by the ordering of event labels α and β and the termination operator Δ. These labels represent instantaneously occurring events such as the communication of a change in voltage or a packet of charge, depending on interpretation. The introduction of expression identifiers and the auxiliary definition operator ⇐ allows recursive definitions of the form

$$A \Leftarrow \lambda \; A$$

Capitals are used to name CIRCAL expressions while small letters and numbers are used to form label names.

5.1 Modelling Continuous Behaviour

The CIRCAL framework is used to interpret continuous circuit behaviour discretely. Many different interpretations of a change in voltage on device ports may be made but for the purpose of this paper changing voltage is modelled by two events; the instantaneous rise in voltage from zero to one and the instantaneous fall in voltage from one to zero. It should be noted that more accurate approximations could be obtained by modelling voltage and charge levels using many more values within the CIRCAL framework.

The simplest device to communicate voltage change is a wire. A change in input propagates through to an output change after some unspecified interval of time. A wire models four distinct events; a rise or fall in input voltage and the rise or fall of the output voltage. Four event ports are then used, one for each event.

A Directional Wire

The wire W has a sort $\{\alpha 0, \alpha 1, \beta 0, \beta 1\}$ and is pictured as

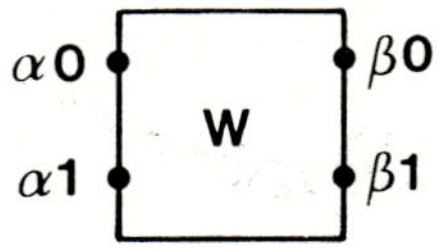

Figure 6

Each physical connection point is modelled by two CIRCAL event ports representing the two possible changes in voltage, rising or falling. For an n-valued logic interpreting the inherently continuous device behaviour we would have n event ports for every physical connection port.

Wire behaviour is given by

$$
\begin{aligned}
W00 &\Leftarrow \alpha 1 \; W10 \\
W10 &\Leftarrow \beta 1 \; W11 + (\alpha 0 \; \beta 1) \; W01 \\
W11 &\Leftarrow \alpha 0 \; W01 \\
W01 &\Leftarrow \beta 0 \; W00 + (\alpha 1 \; \beta 0) \; W10
\end{aligned}
$$

W00 represents a wire with zero voltage on both ends and we use the identifying name to indicate the state of the device. A change in voltage at one end, given by event $\alpha 1$, propagates through the wire causing event $\beta 1$, which may occur on its own or simultaneously with a further voltage change on α, depending on how the environment interacts with the wire. This choice of possible actions is represented using the + operator.

Simultaneity of events is represented by bracketing labels together as with $(\alpha 0 \; \beta 1)$. This composite label represents the α line falling to zero simultaneously with the β line rising to one.

If no voltage change arrives at a wire then it will remain in state W00 or W11 depending on the last input change. In this example voltage change propagates in one direction but a bidirectional wire may also be modelled.

Universal time is modelled by a sequence of instantaneously occurring timing signals or "ticks". When modelling a discrete signal the continuous time axis is broken down into equidistant steps with the signal being sampled at each tick point.

A Timed Wire

If we wish to model the actual rather than the relative occurrence of events on a wire we use five ports, two ports for each physical end of the wire with the fifth being used for timing.

TW of sort $\{\alpha 0, \alpha 1, \beta 0, \beta 1, t\}$ may then be pictured by

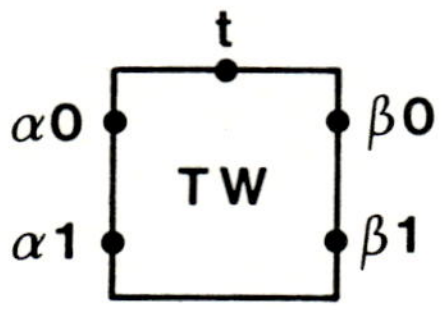

Figure 7

The behaviour of a timed wire is such that input and output only occur on a tick from an abstract timer. This timer represents the passage of time by a sequence of instantaneously occurring timing ticks. The input signal change takes a predetermined number of ticks before it is output; the propagation delay. As the passage of time never halts our wire is defined to accept ticks in the absence of input or output events.

A wire with a one unit propagation delay is defined by

$$
\begin{aligned}
TW00 &\Leftarrow (\alpha 1 \; t) \; TW10 + t \; TW00 \\
TW10 &\Leftarrow (\beta 1 \; t) \; TW11 + (\alpha 0 \; \beta 1 \; t) \; TW01 \\
TW11 &\Leftarrow (\alpha 0 \; t) \; TW01 + t \; TW11 \\
TW01 &\Leftarrow (\beta 0 \; t) \; TW00 + (\alpha 1 \; \beta 0 \; t) \; TW10
\end{aligned}
$$

The abstract timer may itself be modelled as an infinite sequence of ticks or pulses and is then connected to all the t ports of the components being modelled. Timer T of sort $\{t\}$ has behaviour

$$T \Leftarrow t \; T$$

5.2 A Composite Device

An auxiliary relabelling operator may be used to generate instances of a device from some generic description. The postfixed operator $\{\lambda/\mu\}$ recursively replaces all μ labels by λ in the expression it is applied to.

Taking the previous definition of timed wire TW we generate two further wire definitions V and Y which have identical behaviour up to a renaming of labels:

$$V00 \Leftarrow TW00\{\gamma 0/\beta 0\}\{\gamma 1/\beta 1\}$$
$$Y00 \Leftarrow TW00\{\gamma 0/\alpha 0\}\{\gamma 1/\alpha 1\}$$

The two wires are then connected as in Figure 8.

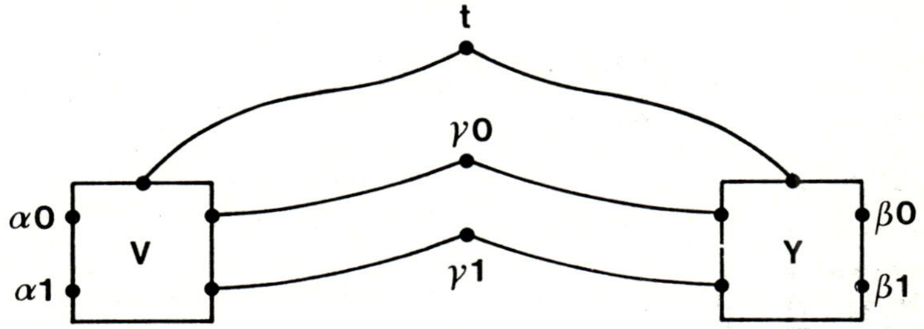

Figure 8

Expressions V and Y are composed using the dot operator to produce the CIRCAL expression which models the two components "running" together and interacting on their $\gamma 0$,$\gamma 1$ and t ports. The recursive dot operator is defined in [4] and [7] but we illustrate its effect using this example.

The CIRCAL model is based on the occurrence of events and it is the synchronous occurrence of two similarly named events belonging to two distinct components which causes them to interact. In the above a $\gamma 1$ event will occur in the composite system only if both the components are able to perform a $\gamma 1$ event at the same point in time. If one component wishes to perform a multiple event such as ($\gamma 1$ t) then this event can only occur in the system if the other component also wishes to perform the same multiple event.

Applying the dot operator to two one-delay wires in a stable zero state results in the following expression

$$V00 \cdot Y00 = (\alpha 1\ t)\ [V10 \cdot Y00] + t[V00 \cdot Y00]$$

Two events may occur; an externally generated $\alpha 1$ event occurs simultaneously with a clock tick or in the absence of an external stimulus an internal t synchronisation occurs on its own leaving both components unchanged. For a stable zero wire the only external stimulus it can receive is when the voltage changes to one.

$$V10 \cdot Y00 = (\gamma 1\ t)\ [V11 \cdot Y10] + (\alpha 0\ \gamma 1\ t)[V01 \cdot Y10]$$

Again two (multiple) events can occur; in both cases the voltage change on the V device has propagated through and is communicated to the Y device by a $\gamma 1$ synchronisation occurring simultaneously with a t synchronisation. If the environment simultaneously generates an α event the ($\alpha 0$ $\gamma 1$ t) will occur.

$$V11 \cdot Y10 = (\beta 1\ t)\ [V11 \cdot Y11] + (\alpha 0\ \beta 1\ t)[V01 \cdot Y11]$$

Here the voltage change propagates out through the β port simultaneously with a tick or with a tick and a further input change. The other states evolving from V00•Y00 have their behaviour established in a similar way.

The passage of time may be modelled at various degrees of granularity with the abstraction operator being used to go from a detailed timing representation to a more approximate representation. System timing using various clocking schema can also be naturally modelled and integrated with universal time. A two-phase clock used to synchronise the behaviour of an equality device is presented at the end of Section 6.

5.3 The Specification Proof Technique

A specification of the intended behaviour of a two-delay device may readily be written in CIRCAL. Specification W200 requires two time units to propogate an input voltage change to an output change. One of the most significant uses of the CIRCAL model is to be able to prove that a constructed device meets its specification and we use this simple example to outline how such a proof proceeds.

The device whose behaviour we wish to examine is hierarchically modelled using the dot operator, mirroring the way in which its components are composed to produce the complete design.

The dot operator is applied to expand the composite wire expression

$$V00 \cdot Y00$$

An abstraction operator is then used to "internalise" the $\gamma 0$ and $\gamma 1$ lines through which component devices V00 and Y00 interact. Applying the abstraction operator removes reference to the γ events

$$[V00 \cdot Y00] - \gamma 0 - \gamma 1$$

Abstraction may also introduce non-determinism, though not in this example, and this property is presented in [7].

CIRCAL axioms such as the associativity and commutativity of + are used to manipulate the expanded expression to determine whether it is syntactically equal to the two delay specification W200, up to recursion.

An acceptance semantics [7] has been used to establish the consistency of the CIRCAL axioms; this allows equivalence proofs to be performed by syntactic manipulation without recourse to the underlying semantics. A simple inductive argument is then used to determine whether the following equality holds

$$[V00 \cdot Y00] - \gamma 0 - \gamma 1 = W200$$

This and other analytical techniques which utilise properties of the CIRCAL model are presented in [8].

The constructive and abstraction features of CIRCAL permit the hierarchical modelling and proof of complex devices. The smallest components of these devices are modelled and verified correct with the proof technique being applied to successively larger objects until a final proof of the whole device is constructed. Using a model related to CIRCAL Gordon [1] demonstrates the applicability of this bottom-up verification technique using realistically sized examples. The following chapter illustrates a different approach with the integration of our specification proof technique into a top-down design methodology.

6. AN EXERCISE IN VERIFIABLE DESIGN

We wish to incrementally design and verify a device which accepts two 8-bit buses as input and produces a 1 on the output if the inputs are identical, or a 0 if they are not. This example is taken from [12] where it is used as a benchmark to examine how a number of commercially available Computer Aided Engineering workstations support modelling using behavioural languages. This exercise only required the device to be modelled and simulated at the top-most level while we are concerned here with designing the equality device down to the gate level and verifying the design as it evolves.

We take an initial specification of the device, produce a design at a level of information content one lower than the specification then verify that the design meets the specification. The implementation components then become the specifications to be implemented and verified at the next lower level of design. This process proceeds down three levels of detail until we reach the gate level where it is assumed that the primitive components with the appropriate behaviour can be taken from a catalogue of component/behaviour pairs. To aid presentation of the verifiable design technique we avoid considering timing issues in detail but we do illustrate how timing may be included in a design with verification of both timing and behavioural features occurring as the design evolves.

6.1 CIRCAL as a Design Language

Early in this paper we introduced the notion of a spectrum of design and description languages while avoiding giving details of what the proposed design languages should look like except to say that they should capture spatial properties such as connectivity and layout. When performing verifiable design the design languages must also capture the behaviour of the device under construction. In this example of verifiable design we use CIRCAL itself as a design language with the CIRCAL notion of labelled ports, sort, port linkages and hiding being used to capture the topology of a design. With a slight extension, illustrated in [5], CIRCAL may also be used to capture the geometrical features of a mask-level design.

Using CIRCAL as both the behavioural model and the design language removes the need for establishing the meaning of a design by mapping from the syntax into the semantics, as illustrated by Figure 2. The behavioural properties of CIRCAL are used directly for design verification. It should also be noted that CIRCAL is used as the single design language bridging the spectrum of design levels from the abstract, functional level down to the gate level.

6.2 Equality Device Specification

The composite identifier E(J,K) is used to name a family of CIRCAL expressions with the variables J and K ranging over the set $\{0,1\}^8$, the domain of boolean 8-tuples. This family of expressions, one expression for each pair of parameters J and K, specifies the intended behaviour of the equality device we are designing and which is informally described above.

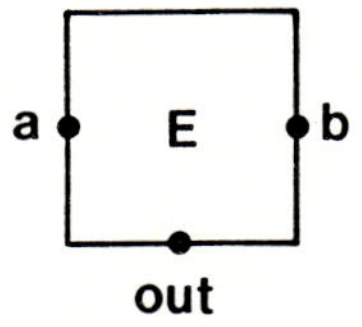

Figure 9

Figure 9 represents the connectivity properties of the device where each of the physical ports labelled by *a* and *b* corresponds to 2^8 CIRCAL event ports with one event port used for each 8-tuple value which may be communicated. Similarly the *out* port corresponds to two CIRCAL event ports; out0 and out1.

Definition : Equality device with sort $\{out0, out1\} \cup \{aX, bX \mid X \in \{0,1\}^8\}$

$$
\begin{aligned}
E(J,J) \Leftarrow\ & \sum_{X \neq J} (aX\ out0)\ \ E(X,J) \\
& + \sum_{Y \neq J} (bY\ out0)\ \ E(J,Y) \\
& + \sum_{\substack{X \neq Y \\ X,Y \neq J}} (aX\ bY\ out0)\ \ E(X,Y) \\
& + \sum_{X \neq J} (aX\ bX)\ \ E(X,X)
\end{aligned}
$$

$$
\begin{aligned}
E(J,K) \Leftarrow\ & \sum_{X \neq K} aX\ \ E(X,K) \\
& + \sum (aK\ out1)\ \ E(K,K) \\
& + \sum_{Y \neq J} bY\ \ E(J,Y) \\
& + \sum (bJ\ out1)\ \ E(J,J) \\
& + \sum_{\substack{X \neq Y \\ X \neq J, K \neq Y}} (aX\ bY)\ \ E(X,Y) \\
& + \sum_{X \neq J,K} (aX\ bX\ out1)\ \ E(X,X)
\end{aligned}
$$

where $J \neq K$

The first clause of the above definition specifies the desired behaviour of the equality device when both inputs take the same value and therefore the output is set at 1. Any change to either of the inputs which results in them becoming non-equal causes the output to change to 0. The second change of the definition assumes that different inputs have been received into the device and the output is then set at 0. Any input change which causes them to become identical results in a 1 appearing on the output via event out1.

The summation notation allows a sum of terms to be produced ranging over the free variable. If there is also a predicate then the summation will range over the free variable such that the predicate holds. For instance

$$\sum_{X \neq K} aX\ E(X,K) \quad \text{is shorthand for}$$

$$aX_1\ E(X_1,K) + aX_2\ E(X_2,K) + \ldots + aX_n\ E(X_n,K)$$

where $X_i \in \{0,1\}^8$ and $X_i \neq K$ for some $K \in \{0,1\}^8$.

The above definition assumes no propagation delay between an output event and the input change which causes it. That is, output events occur simultaneously with the inputs which cause them. This specification is hardly realistic since any implementation will require some propagation delay between input and output.

The following specification assumes that output events occur after the input events which cause them and before any further input changes. No specific timing attributes are made in the informal specification of the device given in [12] or in our CIRCAL definition below. At the conclusion of this section we demonstrate how timing features may be included in the design.

Definition : Equality device with propagation delay.

$$EQ(J,J) \Leftarrow \sum_{X \neq J} aX \;\; out0 \;\; EQ(X,J)$$

$$+\sum_{Y \neq J} bY \;\; out0 \;\; EQ(J,Y)$$

$$+\sum_{\substack{X \neq Y \\ X,Y \neq J}} (aX\ bY) \;\; out0 \;\; EQ(X,Y)$$

$$+\sum_{X \neq J} (aX\ bX) \;\; EQ(X,X)$$

$$EQ(J,K) \Leftarrow \sum_{X \neq K} aX \;\; EQ(X,K)$$

$$+\sum aK \;\; out1 \;\; EQ(K,K) + \sum_{Y \neq J} bY \;\; EQ(J,Y)$$

$$+\sum bJ \;\; out1 \;\; EQ(J,J)$$

$$+\sum_{\substack{X \neq Y \\ X \neq J \\ Y \neq K}} (aX\ bY) \;\; EQ(X,Y) + \sum_{X \neq J,K} (aX\ bX) \;\; out1 \;\; EQ(X,X)$$

where $J \neq K$. $EQ(J,K)$ has the same sort as $E(J,K)$

6.3 Design at Level One

The first level of design and its verification are given in detail to illustrate the verifiable design technique. Subsequent levels of design and their verification are presented informally.

An obvious way of implementing the equality test on the two incoming 8-bit buses is to use eight comparator cells, each of which receive two bit inputs. A 1 is produced by a comparator cell when the inputs change to the same boolean value and a 0 is produced when any input change causes the two inputs to become different. The eight outputs will then be ended together to give the output of the complete device. This first level of design is pictured in Figure 10. Note that each of the physical ports and communication lines corresponds to two CIRCAL event ports or lines; one for a change to 0 and one for a change to 1.

Definition : Comparator Cell with sort {a0,a1,b0,b1,i0,i1}

$$\begin{aligned} C(0,0) \Leftarrow\; & (a1\ i0)\ \ C(1,0) \\ + & (b1\ i0)\ \ C(0,1) \\ + & (a1\ b1)\ \ C(1,1) \end{aligned} \qquad C(1,1) \text{ by symmetry}$$

$$\begin{aligned} C(0,1) \Leftarrow\; & (a1\ i1)\ \ C(1,1) \\ + & (b0\ i1)\ \ C(0,0) \\ + & (a1\ b0)\ \ C(1,0) \end{aligned} \qquad C(1,0) \text{ by symmetry}$$

This definition of a generic comparator cell C assumes no propagation delay between an output-causing input and the output itself. This may seem an unnatural way to model a device whose implementation must involve some form of propagation delay. It is unrealistic and is used for illustrative purposes only, to expedite the presentation of the ideas behind verifiable design. We will indicate later how to include timing properties in a design using CIRCAL, one of whose features is the ability to naturally represent and verify timing issues as illustrated in [7].

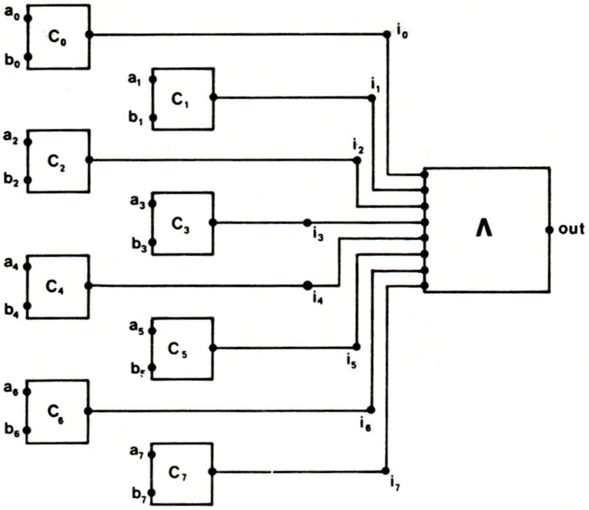

Figure 10

The eight instances of the generic comparator cell are produced using the auxiliary relabelling operator $\{\delta/\gamma\}$ which when post-fixed to a CIRCAL expression replaces all γ labels by δ labels. This relabelling operation may be derived from the primitive CIRCAL operations.

Definition : Cell C_j of sort $\{a_j0, a_j1, b_j0, b_j1, i_j0, i_j1\}$

$$C_j(m,n) \Leftarrow C(m,n)\{a_j/a\}\{b_j/b\}\{i_j/i\}$$

where $m,n \in \{0,1\}$.

The eight-input AND component remains to be defined. As with the original specification of the equality device it is notationally simpler to treat the AND as having one input line taking 8-tuples of boolean values. This approach has the added advantage that the behaviour of the component appears obvious without being hidden in a cloud of notation.

Definition : AND(S) of sort $\{out0, out1\} \cup \{inT \mid T \in \{0,1\}^8\}$ where $S \in \{0,1\}^8$

$$AND(<11\ldots1>) \Leftarrow \sum_{T \neq <11\ldots.1>} inT\ out0\ AND(T)$$

$$AND(S) \Leftarrow in<11\ldots.1>\ out1\ AND(<11\ldots.1>) + \sum_{\substack{T\neq<11\ldots.1> \\ T\neq S}} inT\ AND(T) \qquad \text{where } S\neq<11\ldots.1>$$

The AND component implements an 8-input and device with an unspecified propagation delay before the output event occurs and before further inputs can be received.

A type converter is composed with AND to convert the eight single bit inputs arriving from the eight comparator cells into a single 8-tuple input giving component Λ. Type converters are not implemented and are used solely as an aid to design.

Definition : Type converter $CON_c(X)$ of sort $\{cK \mid K \in \{0,1\}^8\} \cup \{c_i 0, c_i 1 \mid 0 \leq i \leq 7\}$ where $X \in \{0,1\}$

$$CON_c(X) \Leftarrow \sum_{X \neq Y} [\prod_{x_i \neq y_i} (cY \; c_i y_i) \; CON_c(Y)]$$

This abstract component is bidirectional as it converts a change on the bus to changes on the individual bit lines, or vice-versa, without delay. A bus change requires at least one of the bit lines to change.

Note that:

$$X = \langle x_0, \ldots, x_7 \rangle \qquad Y = \langle y_0, \ldots, y_7 \rangle$$

$$\text{and} \quad X \neq Y \Longrightarrow \exists i \, . \, x_i \neq y_i \, .$$

The $\prod$ label operator allows us to construct families of composite labels each satisfying some predicate without having to list them exhaustively. In the above definition all labels $c_i y_i$ where $x_i \neq y_i$ will be included in the composite label by $\prod$.

The $\prod$ label operator is such that

$$\prod_{x_i \neq y_i} (cY \; c_i y_i) \; CON_c(Y) = (cY \; c_m y_m \ldots c_n y_n) \; CON_c(Y) \text{ for all } i \in \{m, \ldots, n\}$$
such that $x_i \neq y_i$

Definition : $CON_{in}(X)$ of sort $\{inK \mid K \; \{0,1\}^8\} \cup \{i_j 0, i_j 1 \mid 0 \leq j \leq 7\}$ where $X \in \{0,1\}^8$.

$$CON_{in}(X) \Leftarrow CON_x(X) \; \{i_j/c_j \mid 0 \leq j \leq 7\} \; \{in/c\}$$

$CON_{in}(X)$ is pictured by Figure 11 where the physical port labelled in passes tuples of eight booleans while each of the ports labelled by i are used to pass a single boolean value when an event occurs. Thus the in port corresponds to 2^8 event ports with every i_j port corresponding to 2 event ports.

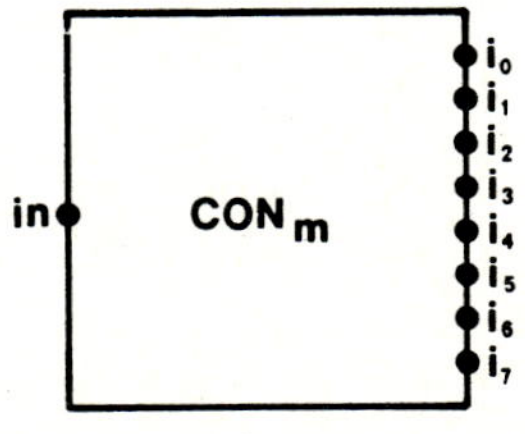

Figure 11

Definition : $\Lambda(X)$ of sort $\{out0,out1\} \cup \{i_j 0, i_j 1 \mid 0 \le j \le 7\}$ where $X \in \{0,1\}^8$

$$\Lambda(X) \Leftarrow [CON_{in}(X) \cdot AND(X)] - in$$

Our specification EQ assumes two inputs each taking 8-tuples of booleans where in fact our implementation uses eight individual boolean inputs for each specification input. Two abstract type conversion devices need to be attached to our implementation to correct this type inconsistency. While we could have converted specification EQ to take individual bit lines as input it is more natural to do the type conversion on the implementation rather than the specification.

Using the definition of converters already given our design at level one is then denoted by

Definition : Level one design $L1(J,K)$ of sort $\{out0,out1\} \cup \{a_K, b_K \mid K \in \{0,1\}^8\}$ with $J,K \in \{0,1\}^8$

$$L1(J,K) \Leftarrow \left[CELLS(J,K) \cdot \Lambda(X)\right] - i_0 - i_1 \ldots - i_7$$

$$X = \langle j_0 \wedge k_0, j_1 \wedge k_1, \ldots, j_7 \wedge k_7 \rangle$$

where

CELLS (J,K) $\Leftarrow$

$$\left[\prod_{0<i<7} C_i(j_i,k_i) \cdot CONV_a \cdot CONV_b\right] - a_0 - a_1 \ldots - a_7 - b_0 - b_1 \ldots - b_7$$

Note that a product operation over the dot operator is used in the above definition. It is defined as

Definition :

$$\prod_{x<i<y} P_i \Leftarrow P_x \cdot P_{x+1} \cdot \ldots \cdot P_y$$

6.4 Proof of Level One Design

We must now verify that our first level of design meets the specification. That is, we wish to prove that repeated application of the dot operator to the eight comparator cells and the type conversion components is then followed by application of the abstraction operator. Remember that the application of the dot operator models the interaction between the physical components as they evolve through time. The abstraction operator is used here to hide the a_i and b_i bit lines so leaving the type-converted a and b ports for the input of 8-tuples.

$$CELL(J,K) = \left[\prod_{0\le i\le 7} C_i(j_i,k_i) \cdot CONV_a \cdot CONV_b\right] - a_0 - a_1 - \ldots - b_7 \quad \text{by definition}$$

$$= \sum_{X \ne J} \left[\prod_{\substack{x_i \ne k_i \\ x_h = k_h}} (aX\ i_i 0\ i_h 1)\ CELLS(X,K)\right]$$

$$+ \ldots$$

$$+ \sum_{Y \neq K} \left[\mathop{\Box}_{\substack{y_i \neq j_i \\ y_h = j_h}} (bY\ i_i 0\ i_h 1)\ \mathrm{CELLS}(J,Y) \right]$$

$$+ \sum_{\substack{X \neq J \\ Y \neq K}} \left[\mathop{\Box}_{\substack{x_i \neq y_i \\ j_i = k_i \\ x_h = y_h \\ j_h \neq k_h}} (aX\ bY\ i_i 0\ i_h 1)\mathrm{CELLS}(X,Y) \right]$$

by definition of composition, abstraction and CELLS.

A CIRCAL system which mechanises the application of the dot and abstraction operators has been produced by Traub [14]. This system, together with the use of the shorthand constructs Σ, $\Box$ and Π, greatly aids the task of performing proofs in the CIRCAL model. They buffer the user both from the inherently complex descriptive notation and the tedious symbolic manipulation exhibited in the above expansion.

The behaviour of $\Lambda(X)$ is formulated using the AND component, which has a single 8-tuple input, and a type converter which splits up the input into eight bit lines.

Application of the dot and abstraction operators gives the following expansion:

$$\Lambda(X) = [\mathrm{CON}_{in}(X) \bullet \mathrm{AND}(X)] - in \quad \text{by definition}$$

$$= \sum_{Y \neq <11...1>} \left[\mathop{\Box}_{x_i \neq y_i} (i_i\ y_i)\ \Lambda(Y) \right]$$

$$+ \sum_{Y = <11...1>} \left[\mathop{\Box}_{x_i = 0} (i_i\ 1)\quad out1\quad \Lambda(<11...1>) \right]$$

by definition of $\Lambda(X)$ where $X \neq <11...1>$

$$\Lambda(<11...1>) = \sum_{Y \neq <11...1>} \left[\mathop{\Box}_{y_i = 0} (i_i\ 0)\ out0\ \Lambda(Y) \right]$$

by definition of Λ .

Composing CELLS and Λ using the dot operator and the above expressions followed by hiding the internalised i and j lines using the abstraction operator produces the following:

$$L1(J,K) = [\mathrm{CELLS}(J,K) \bullet \Lambda(J \wedge K)] - i_0 - i_1 \ldots - i_7 \quad \text{by definition where } J \neq K$$

$$= \sum_{X=K} aX\ \ out1\ \ L1(X,X) \ + \sum_{X \neq K} aX\ \ L1(X,K)$$

$$+ \sum_{Y=J} bY\ \ out1\ \ L1(Y,Y) \ + \sum_{Y \neq J} bY\ \ L1(J,Y)$$

$$+ \sum_{X=Y} (aX\ bY)\ L1(X,Y) + \sum (aX\ bX)\ out1\ L1(X,X)$$

$$L1(J,J) = [CELLS(J,J) \cdot \Lambda(<11\ldots1>)] - i_0 \ldots - i_7 \quad \text{by definition}$$

$$= \sum aX \ out0 \ L1(X,J)$$

$$+ \sum bY \ out0 \ L1(J,Y)$$

$$+ \sum_{X \neq Y} (aX \ bY) \ out0 \ L1(X,Y)$$

$$+ \sum (aX \ bX) \ L1(X,X)$$

We are now in a position to examine the expressions denoted by EQ(J,J) and L1(J,J) and EQ(J,K) and L1(J,K), J ≠ K. These expressions are identical up to the recursive use of their identifiers and a trivial inductive argument completes the proof that the level-one design meets the specification. Note that much of the above proof concerned itself with the type conversion and not the design itself. This overhead would not be significant in an automated verification system such as the CIRCAL System [14].

6.5 Design at Level Two

At this level we are required to design the individual comparator cells together with the eight-input AND device.

A comparator cell may be implemented as an exclusive OR gate in series with an inverter, as pictured in Figure 11.

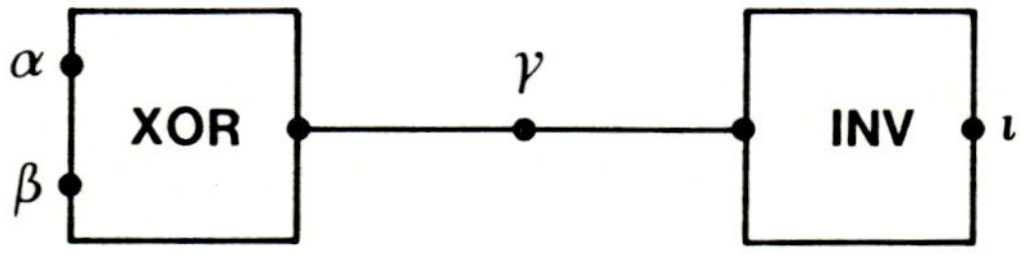

Figure 12

CIRCAL expressions denoting the behaviour of these two components may be defined directly. Examples of various interpretations of the behaviour of NOR, NAND and inverter gates are given in [4],[6] and [14].

This cell design is then verified correct if we can show that:

$$C(a,b) = [XOR(a,b) \cdot INV(a \underline{\vee} b)] - \gamma$$

where $\underline{\vee}$ is the boolean exclusive or operation and $a,b \in \{0,1\}$.

The eight-input AND device may be implemented in a number of ways. Consider a tree of two-input AND gates parameterised on their inputs and interlinked as in Figure 13.

Each AND gate has its behaviour given by a suitable relabelling of a generic two-input AND. Composition and abstraction are again used to establish the behaviour of the design and the eight-input and device is verified correct if the following equality holds:

$$\Lambda(X) = \Big[A(x_0,x_1) \cdot A(x_2,x_3) \cdot A(x_4,x_5) \cdot A(x_6,x_7) \cdot A(x_0 \wedge x_1, x_2 \wedge x_3) \cdot A(x_4 \wedge x_5, x_6 \wedge x_7) \cdot A(x_0 \wedge x_1 \wedge x_2 \wedge x_3, x_4 \wedge x_5 \wedge x_6 \wedge x_7) \Big] - j_0 - j_1 - j_2 - j_3 - k_0 - k_1$$

where $X = \langle x_0, x_1 \ldots . x_7 \rangle$.

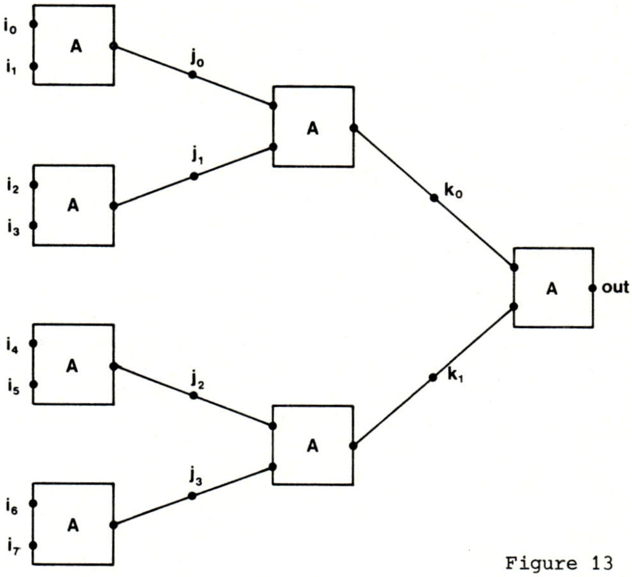

Figure 13

If we are continuing the design process the XOR and AND gates may be implemented using a catalogue of basic components such as inverters, NANDS and NORS. The components have their associated behaviour given in the catalogue and this is used to verify the final step in the design process. The design is now complete and as every design step is verified then the constructed equality device has also been proved correct. Should any step in this iterative process establish that a design fails to meet its specification then a redesign is required with the designer backtracking up some number of design levels.

6.6 Timing as a Design Issue

In the design of the equality device little concern was shown for timing issues such as propagation delay and system clock synchronisation. The reason that such crucially important design issues are avoided is twofold; the informal device specification given in [12] avoids all mention of any intended timing properties and, in any case, we were concerned with presenting the ideas behind the verifiable design technique with the minimum notational complexity.

The above design assumed some unspecified propagation delay across the complete device with an output event occurring after the input events which caused it. A more detailed design involving the verification of timing issues is established if actual propagation delays are explicitly expressed in the behaviour of components at every level of design.

The passage of time may be naturally and accurately represented in CIRCAL by running an abstract (not physical) time line from a universal clock to all components. Continuous time is divided into a sequence of discrete "ticks" with all events occurring simultaneously with such a tick. Any grain of time may be modelled and rigorous techniques exist for moving from a detailed representation of time to a coarser granularity. At every level of design the implementation is verified that it satisfies timing attributes together with the other behavioural features. Proof of design correctness is performed exactly as above since all timing information is embedded in the specification and implementation descriptions.

During the design process a constructed device may fail to meet its specification due to timing problems. The implemented components may be redesigned with different timing attributes such as shorter propagation delays, and the verification stage may be repeated. Device descriptions parameterised on their propagation delays are illustrated in [6],[14] where a flip-flop constructed out of parameterised NOR gates is analysed and these allow us to quickly change a description when the propagation delay is altered. When designing at a level where basic building blocks with fixed timing features are taken from a catalogue of name/behaviour pairs, failure to establish design correctness may be limited by the available components. The specifications must themselves be altered and we are then required to backtrack up the design spectrum.

The incompleteness of the informal description of the equality device given in [12] forced a number of decisions on our design. It was assumed quite rightly that some unspecified propagation delay should separate an output event from the input events which caused it. We also specified that the output event should occur before further input events with only one event ever passing through the device at a time. This causality relationship is imposed by the components in the above design and leads to the arbitrary situation where the eight-input AND device has a propagation delay but the comparator cells are required to be delay-free. This problem arises as the restriction of *further input after preceeding output* should be imposed by the environment in which the equality device is used, and not by the device itself. For instance, in a synchronous system the equality device would probably be connected to input and output latches and a system clock. The equality device, two eight-bit input registers, a one-bit output cell and a two-phase system clock are interlinked as in Figure 14.

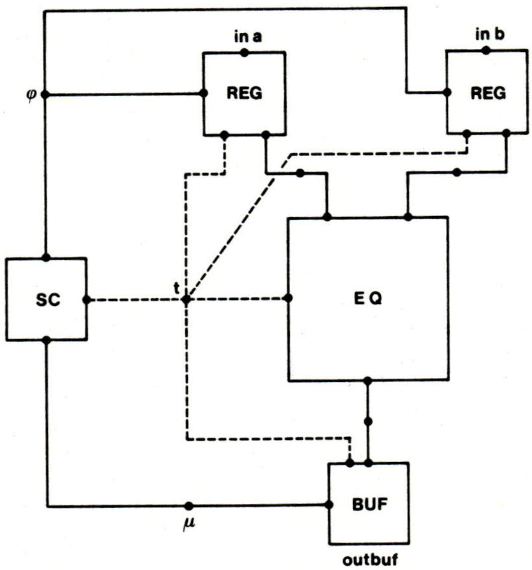

Figure 14

Ø and μ label the two clock lines, one for each phase, with Ø controlling the input latches and μ controlling the output latch. All components of this register-transfer level of design, including the system clock, are synchronised via the abstract t line, so relating the system clock to the propagation delay across the equality device.

A non-overlapping two-phase system clock which restricts the behaviour of the equality device such that only one event may propagate at a time, is defined as:

<u>Definition</u> : SC of sort {Ø0,Ø1,μ0.μ1,t}

$$SC \Leftarrow (\emptyset 1\ t)\ t\ (\emptyset 0\ t)\ t\ (\mu 1\ t)\ t\ (\mu 0\ t)\ t\ SC$$

For the equality device to satisfy this timing regime requires that the total propagation delay across it be less than five time units, given the above grain of time. This requirement would be included in a new timed specification of the equality device and the new design would then be verified as the design proceeds. All components at all levels of design would now have their propagation delays explicitly expressed with a more detailed design being produced than that which we presented above; one in which timing issues are considered and fully verified.

7. DISCUSSION

The production of VLSI designs which can be verified correct by mathematical proof has yet to be achieved but it is a goal which must be reached. This paper identifies verifiable design as <u>the</u> important issue to be addressed by the next generation of design systems. Two techniques are presented, both of which rely on an underlying model of concurrent device behaviour.

Proving the correctness of an automated design procedure allows for the establishment of the correctness of a complete family of designs produced by a silicon compiler from a high-level representation language. This avoids the need for a correctness proof to be constructed for each and every design generated by an automated design system. This technique is presented in [5] where CIRCAL [4,6,7] is used to describe the behaviour of designs at different levels of representation. Related work by Subrahmanyam [13] uses CCS[10] as the basis for a synthesis system producing architectural level descriptions from high-level behavioural specifications. CIRCAL and CCS are related models of concurrency which have evolved from a common root, the Concurrent Process work of Milne and Milner [9].

This paper also presents another technique for verifying design correctness by integrating the design and verification processes. As a design evolves from a high-level, abstract representation through intermediate levels to the mask-level, each design step is verified to establish that the given implementation satisfies the specification at the preceding level of design. CIRCAL is used both as the design language and the behavioural model and an equality device is incrementally designed and verified down to the gate level. This top-down verification may be contrasted to that of Gordon [1] where a microprocessor represented at the register transfer level is verified correct in a bottom-up manner.

The complexity of both design and verification makes it imperative to mechanise these processes as much as possible. The CIRCAL system [14] allows much of the tedium found when performing verification to be removed. An extended version which involves more sophisticated proof techniques is being produced and will be integrated into a prototype design system incorporating many of the ideas presented here.

8. ACKNOWLEDGEMENT

Thanks to Mark Dowson (IST Ltd), Mike Gordon (Cambridge) and Niklas Traub (Edinburgh) for the helpful discussions which contributed to the ideas presented in this paper.

This research was funded in part by the Science and Engineering Research Council and International Computers Ltd.

9. REFERENCES

[1] M Gordon, "Proving a Computer Correct", Report 42, Computer Lab., University of Cambridge, 1984.

[2] C.A.R. Hoare, "Communicating Sequential Processes", Comm. ACM 21,8, 1978.

[3] C. Mead and L. Conway, "Introduction to VLSI Systems", (Addison-Wesley), 1980.

[4] G. Milne, "CIRCAL : A Calculus for Circuit Description", INTEGRATION 1, Nos. 2 and 3, 1983.

[5] G. Milne, "The Correctness of a Simple Silicon Compiler", Proc. 6th Int. Symp. on Computer Hardware Description Languages and their Applications, Pittsburgh, (North-Holland), 1983.

[6] G. Milne, "A Model for Hardware Description and Verification", Proc. 21st Design Automation Conference, Albuquerque, NM (IEEE Computer Scoiety Press) 1984.

[7] G. Milne, "CIRCAL and the Representation of Communication, Concurrency and Time", ACM Trans. on Prog. Lang. and Systems, 7,2, April 1985.

[8] G. Milne, "Simulation and Verification : Related Techniques for Hardware Analysis", Proc. 7th Int. Symp. on Computer Hardware Description Languages and their Applications (CHDL '85), Tokyo, (North-Holland) 1985.

[9] G. Milne and R. Milner, "Concurrent Processes and their Syntax", J.ACM, 26,2, 1979.

[10] R. Milner, "A Calculus of Communicating Systems", LNCS 92, (Springer-Verlag) 1980.

[11] R. Milner, "Calculi for Synchrony and Asynchrony", Report CSR-104-82, Computer Science, University of Edinburgh, 1982.

[12] G. Robson, "Logic Design in Using Behavioural Models", VLSI Design, January 1984.

[13] P.A. Subrahmanyam, "Synthesising VLSI Circuits from Behavioural Specifications : A Very High Level Silicon Compiler and its Theoretical Basis", Proc. VLSI '83, Trondheim, Norway, (North-Holland), 1983.

[14] N. Traub, "A Lisp based CIRCAL environment", Report CSR-152-83, Dept. of Computer Science, University of Edinburgh, 1983.

Formal Aspects of VLSI Design
G.J. Milne and P.A. Subrahmanyam (editors)

DESIGN TRANSFORMATION AND CHIP PLANNING

Robert Milne

Standard Telecommunication Laboratories
London Road
Harlow
Essex
CM17 9NA
United Kingdom

The role of verification in VLSI design is being explored in an experimental system which supports a formal design method for VLSI. The system is based on a logic language which can describe the behaviour, structure and layout of VLSI in ways that vary between abstract specifications and full implementations. Specifications may be transformed interactively or automatically into implementations with the aid of the system. The paper outlines the design method and exposes some of the problems in the construction and use of the system.

1 Introduction

The LTS ('layout and timing for structures') system [2] is a prototype embodying a formal design method. This method regards VLSI design as being the transformation of chip behaviour specifications into mask patterns and manufacturing tests. The transformations may, for instance, make specifications more implementable or alter the structure and layout of designs without changing stated aspects of the behaviour. Some steps in the transformation process are interactive but others are automatic. All the LTS aids used in these steps act on objects in a common data structure which may be viewed in various ways, so we can justly claim that they are 'integrated'.

The LTS system is incomplete but it does cover a significant aspect of design: the refinement of specifications into implementations, especially by introducing structure and layout while preserving chosen parts of the behaviour. This sort of refinement is generally absent from silicon compilers, which in our terms take an implementation in one format and convert it into an implementation in another format. Though we have done some work on silicon compilers we have focussed more of our attention on problems that are less often addressed.

In the system the behaviour is described in a formalism [3] which can be characterised as a higher-order logic programming language. The layout can be viewed in graphical form or as textual annotations that are stored as attributes in an abstract syntax tree (or, actually, an abstract syntax graph).

Despite its being a prototype, the system has been used by friendly users who are prepared to tolerate its lack of user-friendliness. Various chips with up to 70000 transistors have been described in the language and developed with the system, though none has yet made effective use of all the aids in the system.

In this paper we shall discuss both the design method and the design system.

2 The design method

2.1 System design in general

We believe that there is much in common to how hardware and software should be designed, so we shall describe in general terms hypothetical future design methods for either, before specialising to VLSI. These methods rely on more rigorous requirements analysis, the reuse of components and techniques (even in custom design), and the modification of specifications and implementations in ways which can have formal justifications. They are not necessarily top-down: in the course of forming the implementation, the specification itself may be deliberately modified.

The requirements for a design will be analysed and, where practicable, translated into a specification with a formal meaning. The formality of this specification will make it partly susceptible to animation, by, for example, the simulation of its behaviour or the derivation of its properties.

Obtaining the implementation from the specification will involve the transformation of structured objects like programs, proofs and propositions. The transformations will be interactive when automatic aids fail because of computational complexity or the absence of unique useful end results (so a cell generator, say, may not be interactive at all). The interactions will rely on the selection of components and techniques from libraries, the introduction of parameters, and editing. This editing should be constrained by the meanings of the objects being edited (not just by their syntaxes, say), should be motivated by tests predicting its effect, and should be accompanied by conditions stating formally when it is valid. The proofs of these conditions may themselves be carried out by transformations.

The transformations will build up complex patterns of objects which will be kept in step by version control. This should itself work in conjunction with aids to project management which maintain information on the schedule and status of the work by referring to the same objects.

The implementation will differ from the specification in that it is more susceptible to efficient simulation of its behaviour and less susceptible to simple derivation of its properties. During the transformations intermediate versions will arise. After the design is released, a revision will be done by transforming one intermediate version into an implementation; the larger is the revision, the further back in the design history from the original implementation will be the chosen intermediate version.

The sequence of transformations which give rise to an intermediate version may be preserved instead of it; the sequence will itself be an object in the system and may therefore be transformed. In a higher-order system the transformations of transformations should be carried out in the same ways as the transformations of other objects.

2.2 Chip design in particular

To obtain a VLSI design method from the general description above we must at least say:

(a) how the specifications and implementations are expressed;

(b) what constitute the components and techniques;

(c) when the transformations can be automatic.

The specifications are expressed in a logic containing functions because such a language is expressive enough but need not make commitments to particular implementations. The functions can also describe implementations, if there are rules about their interpretations as layout. Some pieces of layout, such as the individual layers in a transistor, do not have behaviours; others, such as pull-up transistors, do have them (in terms of signal strengths) but we prefer to pitch the notion of behaviour at a more abstract level (with signal values but not signal strengths). We also demand that the notion of layout be abstracted so that, for instance, cells have outlines that do not vary between layers. The leaf cells can now be the smallest functions in an implementation that have both behaviour and significance for layout in LTS.

Uses of functions form a hierarchy by appearing within one another. The justification for the term 'leaf cell' in LTS is that the structure of this hierarchy is adopted for both behaviour and layout. In CAD jargon, the LTS system works with 'isomorphic' hierarchies for behaviour and layout but not with 'separated' hierarchies for layout [20], because a cell built from other cells may have a behaviour that depends not only on their behaviour but also on the behaviour of the routes between them. In 3.3 and 3.4 we shall indicate more precisely how the structure of behaviour maps into the structure of layout.

In order that a structure may be described independently of the behaviour (or even the layout) of its constituents the logic actually provides so-called 'higher-order' functions, which take functions as parameters and return functions as results. Such higher-order functions themselves do not have interpretations as layout, though the result of applying them to parameters may do so.

Each leaf cell must be provided as an instance of a reusable component. Some simple components have been codified already (in a CMOS cook book [18], say). Layout can be generated for these automatically, given either a description of their behaviour (in the case of a PLA, ROM or other programmable structure) or a parameterised invocation of objects (such as an ALU). However, in a custom design system there should be no prior commitment to the use of particular components, though there must be an understanding that a component will be given a behaviour specification which can be added to the library and selected in the system.

There is little known about the techniques appropriate to formal (or, indeed, systematic) VLSI design, but work on divide-and-conquer strategies [21] and systolic arrays [15] is a start.

The transformations needed for turning a specification into an implementation therefore include removing higher-order functions, generating the layout for leaf cells, and assembling the layout of the chip from the structure of the whole and the layout of some cells. These transformations can largely be automated, but others cannot. In particular, one transformation which continues to be interactive (albeit supplemented with various automatic aids) is chip planning, which involves adding layout information before the leaf cells are designed. In 4.1 and 4.2 we shall exhibit how chip planning is coupled with behaviour transformation.

In order to make designs as independent of the manufacturing process as possible, the layout information should be in a virtual format (perhaps that of Mulga [24]) till the stage of chip assembly. Inserting the real format during chip assembly then typically involves modifying cells at their boundaries, stretching cells along fracture lines, and routing wires along channels that have already been planned.

2.3 The existing design system in reality

The LTS project is intended to explore the consequences of adopting the design method laid down above, which can be summarised as:

(a) producing a specification;

(b) animating the specification;

(c) transforming the specification into a description having a related meaning and a chip plan;

(d) generating the virtual layout for leaf cells such as PLAs and ROMs;

(e) creating the virtual layout of leaf cells such as ALUs;

(f) relating the predicted performance of the real layout of these cells to the desired performance;

(g) assembling the cells into a chip with real layout;

(h) fabricating the devices;

(i) testing the devices.

So far the LTS project has made available:

(a) a language based on higher-order logic with layout annotations (which in some circumstances can be viewed both textually and graphically);

(b) some simulators;

(c) the 'manipulator' (an interactive transformer that constrains the textual and graphical editing of objects so that users alter the structure and layout of designs without changing the behaviour, sometimes after the system has made suggestions);

(d) the 'expander' (an automatic transformer that makes specifications more implementable by, for example, removing higher-order functions);

(e) some leaf cell generators.

These aids do not form a complete design system: they say nothing about the creation of leaf cells other than programmable ones or about chip assembly and testing. The project has concentrated on the aids that are less commonly considered in CAD and that can form an integrated subsystem of a design system. To provide access to other aids there are also translators from the LTS language to the net list formats of two layout systems, the STC/ITT one and Astra [19]. Because these layout systems pay little heed to behaviour there are no translators in the opposite direction.

Typically once one leaves an integrated subsystem of a design system it is laborious to get back in. If integration is not to extend to the system as a whole the boundaries around integrated subsystems should be drawn at points in the design method where an interactive aid is inappropriate and where little design information needs to be fed back into the integrated subsystems once they have been left. Hence in the LTS system chip planning is done in the manipulator, not in part of a layout system which is not integrated.

A similar point arises in relation to automatic transformers, even when they are integrated. The more advanced they are, the more baffling is their output (except when it can be accompanied by simple and convincing explanations of its evolution). Where possible they should be placed at the end of the design process, so that the designer does not have to look at their results. In the LTS system this is possible with the leaf cell generators and the chip assembler, but not with the expander, which sometimes needs to be followed by interactive chip planning.

3 The language and aids

3.1 The language

The language has the syntax of a programming language but several characteristics of a higher-order logic similar to those of HOL [9] and Veritas [13]. In particular, it treats concurrency by using conjunctions and bidirectionality by using equalities (and formerly by using partial ordering). However, it differs from these other systems in the following respects:

(a) it facilitates the treatment of layout, by letting the user describe aspects of layout, such as the relative positions of 'links' (which are essentially named routes) and the relative positions, sizes and orientations of 'boxes' (which are essentially named cells);

(b) it simplifies the description of timing, by taking the basic entities to be signals mapping time to values (as in a temporal logic or the programming language Lucid [1]), not values themselves, and by insisting that state be referred to through the values of signals at past times;

(c) it provides data typing and name scoping capabilities reminiscent of ML [10], to help to make descriptions comprehensible and consistent;

(d) it includes ranges and arrays among the data type constructors, to handle the structure of VLSI;

(e) it can be simulated almost in its entirety, owing to severe restrictions on where implications can appear and on what is allowed to the left of equality symbols.

We shall devote a paragraph to each of these.

The treatment of layout depends on a relative addressing scheme which refers the positions of links and boxes to corners of the surrounding box; it also provides various other attributes which indicate, for example, the layers occupied by links. An alternative would refer the positions of links and boxes to the positions of other links and boxes. This would be more complicated for the textual syntax and the graphical editing, but it might allow simpler chip assembly by constraint propagation up the hierarchy of boxes.

The description of timing gains much of its simplicity from taking the basic entities to be signals which vary through time, not values independent of time. Thus a latch which takes two inputs, w and v, and which gives as its output the value of v at the most recent time when w was *true* can be expressed as

latch(w,v)
  =**if** w is
    true **then** v,
    false **then** *latch*(*last*(w),*last*(v))
  end

Here *last* is a function which acts as a unit delay by making *last*(x) have as its value at any time the value that x had at the previous time. This unit delay function has been adopted as the basic primitive for timing in the LTS language. A variant (not in the language) with a less obvious dependence on discrete time relies on the expression

when s **was** *m0* **then** *e0*, ..., *mn* **then** *en* **end**

which obtains its value at any time by going back to the most recent time when s matched *mi* for some i (where i lies between 0 and n) and selecting the value of *ei* at that time. This variant would take the language further away from the structure of higher-order logic or temporal logic as it is usually conceived but would be more natural for describing timing hierarchies (in which a unit delay at one level may signify the rise and fall of a clock line at the next level).

The data typing provided leads to more complex type checking than that in ML. In ML the basic entities are independent of time, but the LTS language needs both these entities (for handling structure) and signals which vary through time. Accordingly the basic operators in the language are overloaded: (x,y), for example, might denote either a pair of values or a signal having as its value at any time a pair of values. The overloaded operators are also polymorphic [10], as in ML: the pair of values denoted by (x,y) (perhaps at a given time) may be of a parameterised type. The particular combination of overloading and polymorphism in the LTS language entails having an extra phase of language analysis to distinguish kinds of parameter which are usually sorted out by polymorphic type checking.

The inclusion of ranges and arrays makes full type checking semidecidable (as it is also in programming languages which treat the bounds of arrays as parts of the type but allow arithmetic on array indices). If it were made decidable by restricting the arithmetic to Presburger arithmetic, say, it would be not expressive enough; anyway, a decision procedure can be little more useful than a semidecision procedure, owing to its computational complexity [14]. A semidecision procedure raises the problems of saying what happens before the system gives up and suggesting how the user should continue, which have been wrestled with unsuccessfully by theorem provers besides ours. Probably this is best done in a more general framework, in which the type checking is decidable and the semidecidable aspects of the present types appear among static constraints validated by using the manipulator.

The ability to simulate the language does not require restrictions as severe as those that are actually imposed: at present, for example, some Horn clauses are not permitted. Moreover, we may want specifications which are not implementable even in principle (but which use constructs, such as case expressions, that foster their consistency instead of those, such as implications, that do not). Weakening the restrictions has not yet reached the top of our list of priorities, as it should be accompanied by introducing higher-order functions for structuring theories, like those implicit in Clear [7] and Larch [11].

3.2 The simulators

Much of the LTS language is a functional programming language. This fact is exploited in the early simulators for LTS, which translate LTS into the functional subset of ML and then execute that. Each signal is treated as a mapping from time to values. Doing this has the effect that the time for simulation is often quadratic and sometimes exponential in the number of time steps. For instance, simulating a use of the recursive definition

latch(w,v)
  =**if** w is
    true **then** v,
    false **then** *last*(*latch*(w,v))
  end

may involve calculating all the previous values of *latch*(w,v) every time that a new one is calculated.

Subsequent simulators make the obvious optimisation of storing the values of a signal at certain earlier time steps and updating them as required. However, doing this involves restricting how much of the LTS language can be simulated: though uses of the definition of *latch*(w,v) above can be simulated, uses of the equivalent definition

latch(w,v)
  =**if** w is
    true **then** v,
    false **then** *latch*(*last*(w),*last*(v))
  end

cannot be simulated, because we do not know independent of w and v how many earlier values must be kept. We use the expander to transform such definitions into ones which can be simulated.

The LTS language is not just a functional programming language: it includes equality predicates and quantifiers. If the simulators were to provide symbolic execution the equality predicate would be treated by unification in the conventional logic programming style. As the simulators act only on actual data, however, assignment can be used instead of unification, as long as only compound names are allowed to the left of equality symbols and errors are produced by assigning to signals values not equal to their existing ones. Of course, the test for equality is undecidable for higher-order values, but so is unification [8].

The other details of the first-order part of the LTS linear-time simulator are quite similar to those described fully for the simulator for Tempura [17], so we need not say more about them.

3.3 The manipulator

At present the manipulator offers three main kinds of editing, all of which are constrained by the meanings of the objects being edited. They are:

(a) editing the behaviour representation (perhaps giving it a different structure and layout) by referring to the textual view of a specification (as in a syntax-directed editor [22]) and applying rewrite and induction rules;

(b) editing the layout information by referring to the graphical view of a specification and applying fairly conventional chip planning operations;

(c) editing the behaviour representation and layout information at the same time by referring to the graphical view of a specification and applying chip planning operations (such as 'introducing a bounding box') which are possible only when the behaviour and layout are integrated.

We shall consider these briefly one by one, before remarking on some aspects of the manipulator interface.

Editing the behaviour representation preserves all the behavioural meaning (in the versions of the system made so far: there is no way of coarsening or refining behavioural meanings. We imposed this discipline to make the system usable by people unfamiliar with the styles of reasoning needed to operate LCF [10] effectively. Doing this has proved to be wrong: several users have demanded manipulations which do not preserve all the behavioural meaning and seem willing to tolerate the consequences (needed by CIP [4], for example).

Editing the layout information involves changing attributes which do not relate to behaviour. The changes can be done by the user, by the system to create suggested chip plans, or by the system to keep the whole design automatically consistent. The system can currently make suggestions about positioning and orienting boxes and pins; for boxes these are like those of Champ [23] and for pins they act to prevent crossovers within and between links. Formerly there were also algorithms for sizing boxes and inserting power-ground combs, on the assumption that the boxes abutted, but these have not yet been made available in the latest version of the system. All these automatic chip planning aids have proved their worth, as they can vastly reduce the amount of layout information which the user must provide but they do not limit the freedom of the user to make critical design decisions.

Editing the behaviour representation and layout information at the same time is typically easy if structure is being removed but difficult if structure is being added. Thus, removing a bounding box by pointing at the layout simply involves unfolding (in the sense of program transformations [6]) a function invocation in the text (provided that the layout positions and orientations in the text are adjusted to relate to an outer box instead of the removed one), because boxes are taken to be function invocations. On the other hand, introducing a bounding box by pointing at the layout does not simply involve folding a function invocation in the text, since the textual positions of names may have no relation to the layout positions of the links and boxes with which they are associated. This difficulty diminishes if we drop all first-order functions other than predicates and restrict our attention to abutting boxes (when their relative addresses can be defined by conjunctions which are marked as 'upwards' or 'rightwards'); however, doing this may just move effort to part of the design process where graphical views are not possible.

Most of the manipulations in the system are presented to the user as invocations of procedures in ML, so the user can construct sequences of transformations which are automatically saved and may be reused. The only exceptions to this rule are some chip planning operations, to which the 'parameters' are passed in response to prompts, not along with the invocation itself. These presentations of how parameters are passed can be reconciled by separating the human interface from the transformation language (in our case, ML), but we have not yet done this.

Even if we did separate the human interface from the transformation language, there would remain an obstacle to reusing sequences of transformations. The operations performed by the user to select nodes in an abstract syntax tree move cursors to a name or along branches of the tree towards an instance of a pattern which the user never specifies. However, a general programmed transformation tends to specify a pattern on which it acts, rather than a name or the branches by which an instance of the pattern can be reached. The discrepancy between these versions of the transformation may not be resolved properly till we know something about the roles of analogy and induction in design.

The manipulator, like any syntax-directed editor, can change not only the context free syntax of a specification but also the context dependent attributes. The system keeps some of these attributes in step with the changes but recalculates others only when they are needed. Keeping the scopes of names in step is straightforward, but keeping the types in step is harder, because in a polymorphic language such as LTS the composition rules (by which the types of the whole synthesise, or are synthesised from, the types of the parts) appear too complex to be expressed efficiently in an attribute grammar. Though types are now maintained automatically this is done not by synthesising attributes but by propagating constraints.

By contrast with types, nets (describing the connections between boxes) are reconstructed only by chip planning operations, which use graphical views of objects. Though manipulations of behaviour which use textual views may invalidate nets, this does not matter till graphical views are wanted.

3.4 The expander

The expander transforms specifications into others that are more implementable. By 'more implementable' here we mean that the names that occur in them can be regarded as individual links and boxes.

To make this more precise, we say that a signal 'does [not] need pins' if the type of its value at any time is constructed without the use of function spaces and without the use of [anything except] the type of the value returned by the equality predicate at any time. Only signals which do need pins can be sent along wires; only signals which do not need pins can arise from equality predicates. Accordingly, by an LTS convention,

(a) an individual link should take no parameters (because such parameters would be inputs to a box) and should denote a signal which does or does not need pins;

(b) an individual box should take no parameters other than signals which do need pins, should mention no signals other than parameters which do need pins, and should have a result that denotes a signal which does or does not need pins.

Signals which do need pins are distinguished sharply from those that do not need pins because the equality predicate in the LTS language is a logical statement, not something with an output. This distinction seems to be helpful to users, but it greatly complicates the chip planning operations. We could eliminate it by dropping all first-order functions other than predicates from the language; doing this would not reduce the power of the language but would reduce its convenience.

Most objects other than individual links or boxes should be treated as generators of structure which need to be 'executed' before layout is obtained; hence the LTS system has the capability to do procedural layout. In particular, the expander repeatedly unfolds invocations of higher-order functions to remove them in favour of the individual links and boxes which constitute them. When the higher-order functions are recursive (as they generally are, in interesting structures) unfolding them might duplicate some of the boxes already present.

The expander avoids such duplication by introducing named copies of the boxes with broad scopes and folding all the duplicates with these named copies.

Some objects which should be regarded as boxes mention signals that are not parameters and that do need pins. The expander may introduce extra parameters to make these objects into boxes, but it will not currently introduce power, ground and clock parameters unless the corresponding signals appear explicitly.

In addition, the expander simplifies expressions and removes redundancy in various ways.

The expanded specification is very often implementable when the unexpanded one is not. However, it is very often quite different from the unexpanded one. Because the expanded specification may still need to be subjected to interactive transformation (particularly chip planning) before the final automatic stages of the design process, the difference between the expanded and unexpanded versions of the specification can be awkward for the user. Even examining the trace of how and why the expander altered the specification may not be greatly illuminating. This situation, in which the output from a program may still perplex the user when its history is traced, can arise elsewhere in the LTS system (in type checking and in testing that function results are defined) and is familiar to users of other system with similar power (including even the Boyer-Moore theorem prover [5]); splitting the program into simpler tactics and providing better interactive facilities may be a better way to deal with it than pressing towards more automation.

3.5 The leaf cell generators

The leaf cell generators currently act on a subset of the LTS language built from case expressions of the form

if *s* **is** *m0* **then** *e0*, ..., *mn* **then** *en* **end**

and logical operations. They create virtual and real formats for CMOS layouts (with clocked pull-up transistors) of PLAs and *nor* matrix gates [16]. When the users so desire, the layouts may be minimised and folded (in the sense of layout optimisations [12]).

The leaf cell generators also also have their own interactive transformation subsystem so that users may decide how best to partition logic among several logic arrays. This subsystem should be merged with the manipulator, where the measures of partitioning that it employs could be put to broader use.

4 Examples

4.1 A latch

By looking at the latch again we can give a simple example of how developing the structure of the behaviour can develop the structure of the layout. The example is unrealistic, in that normally we would expect a latch to be available as a leaf cell in which the structure of the behaviour did not match the structure of the layout. However, the principle of how the transformations are carried out is valid irrespective of the example.

The changes will be described in six stages. The first and second of these would in practice be carried out by the expander. The third and fourth could in principle be carried out by the expander, but currently require sequences of manipulator operations. The fifth at present entails resorting to a text editor; there should be a mode of use of the manipulator in which the operations that it needs are available in a syntax-directed manner. The sixth requires a sequence of manipulator operations which view the description graphically in order to do chip planning.

We start with the definition

latch(w,v)
  =**if** w is
    true **then** v,
    false **then** *latch*(*last*(w),*last*(v))
  end

As we noted above, this cannot be simulated by means of the linear-time simulator. By the same token, it cannot be laid out, as *latch*(w,v) (a latch with inputs w and v) is defined in terms of *latch*(*last*(w),*last*(v)) (a latch with inputs *last*(w) and *last*(v)), so its layout would be an infinite sequence of latches with different inputs. We therefore transform it into

latch(w,v)
  =**if** w is
    true **then** v,
    false **then** *last*(*latch*(w,v))
  end

This definition at least describes *latch*(w,v) in terms of itself, but it still seems to suggest that within the box *latch*(w,v) there is another box, *latch*(w,v). We avoid having this other box by introducing a link x which feeds back within the definition of *latch*(w,v), thereby obtaining

```
latch(w,v)
  =x
    where
      x
        =if w is
            true then v,
            false then last(x)
          end
    end
```

We can identify the case expression in the definition of *latch*(*w*,*v*) with a box, as long as we give the box a name and enough parameters to cater for the three things on which the case expression depends. We therefore introduce a definition of the box and fold it with the case expression. Doing this leads to

```
latch(w,v)
  =x
    where
      x
        =select(w)(v)(last(x)),
      select(w)(v)(u)
        =if w is
            true then v,
            false then u
          end
    end
```

In the LTS system *last*(x) is treated as a signal which is x delayed by one time step, but what we require in an actual implementation of a latch is a buffer for the values which are fed back; as we have chosen to omit signal strengths, we model a buffer by a delay in an appropriate context. Introducing a definition of this buffer and folding it with the use of *last* produces

```
latch(w,v)
  =x
    where
      x
        =select(w)(v)(buffer(x)),
      select(w)(v)(u)
        =if w is
            true then v,
            false then u
          end,
      buffer(x)
        =last(x)
    end
```

We now specialise the functions so that they refer to logical values (which require links comprising one wire each), we provide a size of 50 units by 60 units for the outermost box, and we ensure that *buffer*(x) is treated as a leaf cell; *select*(u)(v)(w) will be automatically turned into a leaf cell because it uses a case expression. The types which we do not enter will be calculated automatically. After we have done this specialisation, the description is

```
logiclatch(w,v)%(50,60)
  =x
    where
      x
        =select(w)(v)(buffer(x)),
      select(w)(v)(u):logic
        =if w is
            true then v,
            false then u
          end,
      buffer(x):logic%leaf
        =last(x)
    end
```

The apparent layout for this consists of the outline of one box, with no visible components. We can, however, add layout information by using chip planning operations which refer to the graphical view of the object. When solid lines outline boxes and dashed lines supply links, the resulting graphical view is

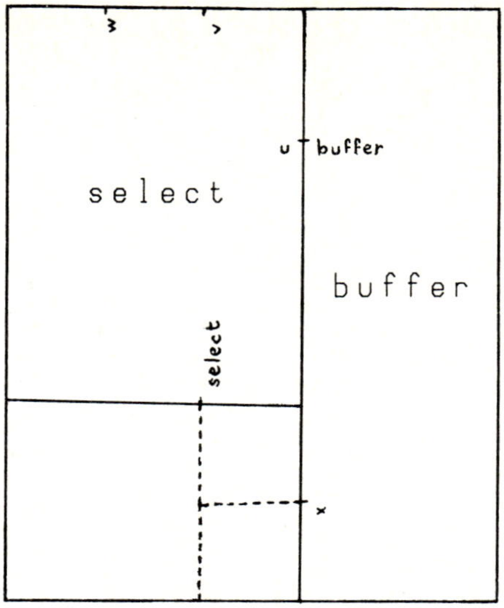

The corresponding textual view describes the relative positions of boxes, links between boxes, and pins on boxes in terms of co-ordinate pairs, vectors of co-ordinate pairs, and side names and co-ordinates (respectively); for instance, it indicates that the output from *logiclatch*(*w*,*v*) emerges 20 units from the left corner of the bottom side. It takes the form

```
logiclatch(w@north 10,v@north 20)@south 20%(50,60)
  =x@@((20,0);(20,20))
    where
      x
        =select(w)(v)(buffer(x@@((30,10);(20,10);(20,20)))%%(30,0))
          %%(0,20),
      select(w@north 10)(v@north 20)(u@east 30):logic@south 20
        %(30,40)
        =if w is
            true then v,
            false then u
          end,
      buffer(x@west 10):logic@west 50%leaf (20,60)
        =last(x)
    end
```

4.2 An array of latches

We shall now extend the treatment of latches to cover an array of latches. We could generalise the discussion of this to arbitrary arrays, but this is best done using predicates (rather than input-output functions) to describe behaviour and we do not wish to introduce yet more notation.

We begin by noting that our definition of *latch*(w,v) does not require that *latch* have a particular type. Accordingly we can specialise it to a version in which v is an array having type $0..n^{*}{>}t$, which indicates that it takes an index in the range $0..n$ and returns a result having type t. Here n and t are allowed to vary, so they appear as parameters of the defining occurrence of *arraylatch*, but they are determined implicitly by the context of an invocation of *arraylatch*, so they do not appear as parameters of the invocation. We therefore enclose them in curly brackets, writing

$$arraylatch\{n,t\}(w,v{:}0..n^{*}{>}t)$$
$$=latch(w,v)$$

We wish to turn the single box *latch*(w,v) into an array of boxes which we can implement. We begin by providing an index, thereby getting

$$arraylatch\{n,t\}(w,v{:}0..n^{*}{>}t)[i{:}0..n]$$
$$=latch(w,v)[i]$$

This still has only one box (albeit with an indexed output). We apply induction to convert it into a definition in which the occurrence of *latch* has a different type and represents an element of an array. This new definition is

$$arraylatch\{n,t\}(w,v{:}0..n^{*}{>}t)[i{:}0..n]$$
$$=latch(w,v[i])$$

We next specialise t to be *logic*, by setting

$$arraylogiclatch\{n\}(w,v{:}0..n^{*}{>}logic)$$
$$=arraylatch(w,v)$$

We can modify the induction mentioned above to show that

$$arraylogiclatch\{n\}(w,v{:}0..n^{*}{>}logic)\ [i{:}0..n]$$
$$=logiclatch(w,v[i])$$

We now wish to indicate the path of the link w, the path of the link $v[i]$, and the position of the box *logiclatch*$(w,v[i])$ (relative to the bottom left corner of *arraylogiclatch*(w,v)). We position the box *logiclatch*$(w,v[i])$ at the bottom of the array and arrange for w and $v[i]$ to be routed into it. To do this we assume that *logiclatch* has width *logiclatchwidth* and depth *logiclatchdepth* and that the inputs to *logiclatch* have insets *logiclatchwinset* and *logiclatchvinset* respectively (relative to the bottom left corner of *logiclatch*); at present the user has to define entities like *logiclatchwidth* and *logiclatchdepth* separately from *logiclatch*, instead of picking them up as attributes of the name *logiclatch*. We must also provide a routing channel having width *gap**2 because *latch*(*w*,*v*) does not provide *w* as an output which could be used by an adjoining latch. Accordingly we form a generic function, covering both behaviour and layout, by writing

```
arraylogiclatch{n}(w,v:0..n*>logic)[i:0..n]
  %(logiclatchwidth*(n+1),logiclatchdepth+gap*2)
  =logiclatch(
      w@@((logiclatchwidth*i+logiclatchwinset,logiclatchdepth);
          (logiclatchwidth*i+logiclatchwinset,logiclatchdepth+gap);
          (logiclatchwinset,logiclatchdepth+gap);
          (logiclatchwinset,logiclatchdepth+gap*2)),
      v[i]@@((logiclatchwidth*i+logiclatchvinset,logiclatchdepth);
          (logiclatchwidth*i+logiclatchvinset,logiclatchdepth+gap*2)))
    %%(logiclatchwidth*i,0)
```

When we want, say, four latches, we give *n* a value and write

```
quadarraylogiclatch(w,v:0..3*>logic)
  %(logiclatchwidth*(3+1),logiclatchdepth+gap*2)
  =arraylogiclatch(w,v)
```

Using the expander on this definition with *logiclatchwidth*, *logiclatchdepth*, *logiclatchwinset* and *logiclatchvinset* having the values implied by *logiclatch*(*w*,*v*) (namely 50, 60, 10 and 20) and with *gap* having the value 10 gives us something which has as its graphical view

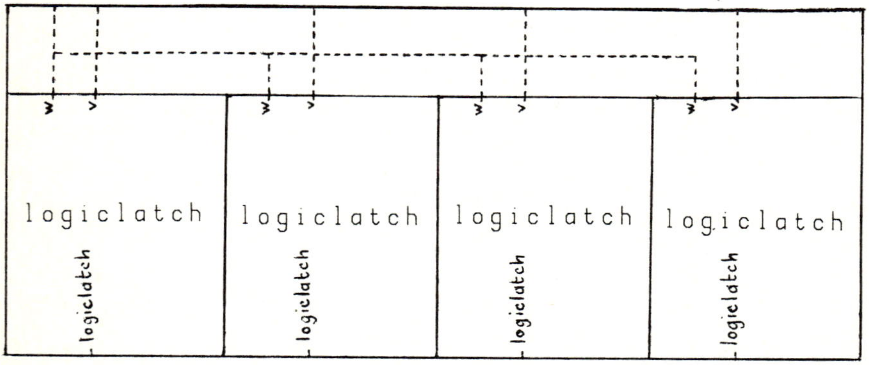

The textual view produced by the expander records that the output from *quadarraylogiclatch*(*w*,*v*) is a vector made from the outputs from four instances of *logiclatch*(*w*,*v*) which have different inputs. It takes the form

```
quadarraylogiclatch(w,v:0..3*>logic)%(200,80)
  =logiclatch0;logiclatch1;logiclatch2;logiclatch3
    where
      logiclatch0
        =logiclatch(
            w@@((10,60);(10,70);(10,70);(10,80)),
            v[0]@@((20,60);(20,80)))
          %%(0,0)
    end
    where
      logiclatch1
        =logiclatch(
            w@@((60,60);(60,70);(10,70);(10,80)),
            v[1]@@((70,60);(70,80)))
          %%(50,0)
    end
    where
      logiclatch2
        =logiclatch(
            w@@((110,60);(110,70);(10,70);(10,80)),
            v[2]@@((120,60);(120,80)))
          %%(100,0)
    end
    where
      logiclatch3
        =logiclatch(
            w@@((160,60);(160,70);(10,70);(10,80)),
            v[3]@@((170,60);(170,80)))
          %%(150,0)
    end
```

5 Conclusion

We are not clear how readily formal methods will be adopted for VLSI design. Most of the users have been enthusiastic about the LTS system. However, they have not yet used it through all the phases of the design process that it covers, partly because of the inefficiency and incompleteness of the prototype and partly because of the limitations in the system mentioned above.

Even if the system were perfect, introducing formal design would be arduous: product developers are reluctant to deviate from methods and aids for which they know both the strengths and the weaknesses. With skilled and motivated users we have still found that the aids which we are most strongly urged to enhance are those such as the simulators which can fit existing methods.

6 Acknowledgements

The LTS system has been created by a team effort. The author of this paper is extremely grateful to the past and present participants in the team at Standard Telecommunication Laboratories, namely Sue Babiker, Rob Belson, Tony Evans, Roger Fleming, Chris George, Howard Gill, Hamid Lesan and Kevin Waite. Their contributions should be the subjects of subsequent papers exploring aspects of the system in greater depth.

7 References

[1] Ashcroft, E.A., and Wadge, W.W., 'Lucid: A Formal System for Writing and Proving Programs', *SIAM Journal of Computing* **5** (1976) 336-354.

[2] Babiker, S.A., Evans, A.J., Fleming, R.A., George, C.W., Milne, R.E., and Waite, K.W., 'A guide to designing with LTS', Internal Technical Memorandum 311.85.5, Standard Telecommunication Laboratories (1985).

[3] Babiker, S.A., Fleming, R.A., and Milne, R.E., 'A tutorial on LTS', Internal Technical Memorandum 311.85.4, Standard Telecommunication Laboratories (1985).

[4] Bauer, F.L., and others, *The Munich Project CIP: Volume 1: The Wide Spectrum Language CIP-L*, Lecture Notes in Computer Science 183 (Springer-Verlag, Berlin, 1985).

[5] Boyer, R.S., and Moore, J.S., *A Computational Logic* (Academic Press, New York, 1979).

[6] Burstall, R.M., and Darlington, J., 'A transformation system for developing recursive programs', *Journal of the ACM* **24** (1977) 44-67.

[7] Burstall, R.M., and Goguen, J.A., 'An Informal Introduction to Clear', in Boyer, R.S., and Moore, J.S., (editors) *The Correctness Problem in Computer Science* (Academic Press, New York, 1981) 185-213.

[8] Goldfarb, W.D., 'The Undecidability of the Second-Order Unification Problem', *Theoretical Computer Science* **13** (1981) 225-230.

[9] Gordon, M.J.C., 'Why higher-order logic is a good formalism for specifying and verifying hardware', Technical Report 77, University of Cambridge Computer Laboratory (1985).

[10] Gordon, M.J.C., Milner, A.J.R.G., and Wadsworth, C.P., *Edinburgh LCF*, Lecture Notes in Computer Science 78 (Springer-Verlag, Berlin, 1979).

[11] Guttag, J.V., Horning, J.J., and Wing, J.M., 'The Larch Family of Specification Languages', *IEEE Software* **2** (1985) 24-36.

[12] Hachtel, G.D, Newton, A.R., and Sangiovanni-Vincentelli, A.L, 'An Algorithm for Optimal PLA Folding', *IEEE Transactions on Computer-Aided Design of Integrated Circuits and Systems* **CAD-1** (1982) 63-77.

[13] Hanna, F.K., and Daeche, N., 'Specification and Verification using Higher-Order Logic', in Koomen, C.J., and Moto-oka, T., (editors) *Computer Hardware Description Languages and their Applications* (North-Holland, Amsterdam, 1985) 418-433.

[14] Hartmanis, J., *Feasible Computations and Provable Complexity Properties* (SIAM, Philadelphia, 1978).

[15] Kung, H.T., 'Why Systolic Architectures?', *IEEE Computer* **15** (1982), 37-46.

[16] Lopez, A.D., and Law, H.F.S., 'A Dense Gate Matrix Layout Style for MOS LSI', *IEEE Journal of Solid-State Circuits* **SC-15** (1980) 736-740.

[17] Moszkowski, B., 'Executing Temporal Logic Programs', Technical Report 71, University of Cambridge Computer Laboratory (1985).

[18] Newkirk, J.A., and Mathews, R.G., *The VLSI designer's library* (Addison-Wesley, Reading, 1983).

[19] Revett, M.C., and Ivey, P.A., 'Astra- A CAD system to support a structured approach to IC design', in Anceau, F., and Aas, E.D., (editors) *VLSI'83* (North-Holland, Amsterdam, 1983) 413-422.

[20] Rowson, J., 'Understanding Hierarchical Design', Ph.D. Thesis, California Institute of Technology (1981).

[21] Smith, D.R., 'Top-Down Synthesis of Divide-and-Conquer Algorithms', *Artificial Intelligence* **27** (1985), 43-96.

[22] Teitelbaum, T., and Reps, T., 'The Cornell Program Synthesizer: A Syntax-Directed Programming Environment', *Communications of the ACM* **24** (1981) 563-573.

[23] Ueda, K., Kitazawa, H., and Harada, I., 'Champ: Chip Floor Plan for Hierarchical Layout Design', *IEEE Transactions on Computer-Aided Design of Integrated Circuits and Systems* **CAD-4** (1985) 12-22.

[24] Weste, N., 'Mulga- An Interactive Symbolic Layout System for the Design of Integrated Circuits', *Bell System Technical Journal* **60** (1981) 823-858.

Formal Aspects of VLSI Design
G.J. Milne and P.A. Subrahmanyam (editors)

Digital Design in a Functional Calculus

Steven D. Johnson

Indiana University
Computer Science Department
Bloomington, Indiana
U.S.A.

A technique is presented for deriving synchronous-system descriptions from recursive function definitions, using correctness-preserving transformations. The use of functional abstraction to address control and communication is illustrated in a small example. The relationship of this technique to conventional digital design methodology is discussed.

Introduction

This paper develops the thesis that functional notation is natural for the description of digital circuits and that functional calculus is an appropriate basis for digital engineering. Standard function transformation techniques are adapted to the problem of digital circuit synthesis. Functional programmers will find familiar transformation methods applied in an unfamiliar setting; digital engineers will recognize a structured design technique cast in a functional notation. This material is intended for the union of these two groups, not the intersection. The presentation is informal but involves disparate formalisms.

The term "synthesis" is used above according to Manna's definition [4]: *...the theory for constructing ...[implementations] that are guaranteed to be correct [with respect to their specifications] and therefore do not require debugging or verification.* The emphasis here is on "theory", not "constructing". Before automation is attempted, it should be shown that an engineering methodology is palatable and pervasive. It must reflect intuitions about the implementation realm because synthesis requires intelligent guidance. The synthesis process should yield correct implementations, not just correct descriptions of implementations—a distinction that has more bearing in hardware than in software. The technique presented here borrows from an established methodology for software engineering and is also related to practical digital design techniques. However, the scope of its applicability in practice has not been demonstrated.

Manna uses "verification" in reference to logical correctness. In hardware, this term encompasses analytic questions concerning the satisfaction of physical design constraints, testability, presence of spurious electrical properties, and so forth. These kinds of problems are not considered in this paper.

The synthesis theory used here is based on transformation, what Darlington and Burstall describe as "an inference system in which the sentences are recursion equations" [2]. A problem is stated in the form of a recursive function definition, or *specification*. The goal is to systematically derive a circuit description using correctness-preserving transformations. Whether the specification itself is right or wrong is immaterial; the object is to obtain a correct circuit relative to the problem statement.

The transformation target is a recursion scheme (or pattern) that characterizes synchronous digital systems. Syntactically, this target is a schematic in linear form; it states the components and the connectivity of the corresponding circuit. The characteristic scheme is restrictive, but it can be mechanically derived from the larger class of *iterative* specifications. In essence, iterative functions are the language of structured digital design.

Transformation of arbitrary specifications to iterative form (also known as "recursion removal") cannot be automated in the most general sense [12]. This problem is central to the use of functional calculus to describe computation and has been studied extensively. The explicit treatment of control through the use of *continuations* is a paradigm for the methodology. An aspect of the implementation is introduced as a functional abstraction. The abstraction is then replaced by an appropriate representation, whose interpretation is added to the specification. This process continues until specification is sufficiently concrete.

One advantage of using functional notation in design is its executability. The engineer can experiment directly in the design language, making the precarious translation to simulation packages and other modeling vehicles, in principle, unnecessary. In software the analogy is to use a modeling vehicle, such as Lisp, to quickly develop, explore, and refine a prototype—or "executable specification"—from which an efficient program derives. This work uses a "lazy" Lisp-like language called Daisy, which permits both functions and data to be defined recursively. The normal-order semantics of a lazy language is preferred because of its adherence to the mathematical meaning of functional notation. *Data recursion* is especially useful for modeling circuits; non-terminating digital behavior is readily represented by non-finite lists, sometimes called *streams*. Each of the functional expressions shown in this paper is a stylized Daisy program, but none is peculiar to it. Laziness and streams are easily added to any language that properly provides for functional values [1, 3].

After a review of notation, a simple model of digital behavior is presented. The main sections illustrate the synthesis technique with an example. The conclusion discusses the relationship to conventional design methods and illustrates the use of functional languages in experimentation.

Functional Specifications

Specifications are built from constants, identifiers, operations, applicative terms (e.g. $f(e_1, \ldots, e_n)$), conditional expressions (e.g. $p \rightarrow e_0, e_1$), and function definitions (e.g. $F(x_1, \ldots, x_n) \Leftarrow e$). A term is *simple* if it contains no defined function symbols. The notation is embellished with syntax for making specifications easier to read. One might specify the *factorial* function as

$$Fac(n) \Leftarrow zero?(n) \rightarrow 1,\ multiply(n,\ Fac(decrement(x)))$$

A functionally equivalent version of *Fac* is given by

$$\begin{array}{l} F(n,1) \\ \quad where \\ \quad F(x,y) \Leftarrow zero?(x) \rightarrow y,\ F(decrement(x), multiply(x,y)). \end{array}$$

This version is *iterative*: each instance—the single instance in this case—of a defined-function symbol is tail-recursive (outermost) [6]. Specifications can have an operational interpretation beyond the functions they define. The iterative version of *Fac* is a better *program* because it executes without a recursion stack. It is essentially a flowchart [7].

Lambda-notation is used to specify and manipulate functional abstractions. The expression $\lambda xy.e$ denotes the function defined by expression e with formal parameters x and y. The form $\lambda(x,y).e$ is used when the formal parameter is a pair. The free identifiers in e are bound lexically, allowing one to build abstractions from the surrounding context.

A Model of Behavior

A digital circuit is a system of non-terminating behaviors, or *signals*, formally modeled as infinite sequences. Considered as a domain,

$$SIGNAL_V = V \times SIGNAL_V.$$

V is a domain of ground values. "$\langle v_0, \langle v_1, \langle v_2, \ldots \rangle\rangle\rangle$" is abbreviated to "$\langle v_0, v_1, v_2, \ldots \rangle$". Signals model the history of values on a conducting path. They are infinite because circuits do not terminate in the sense that programs do. A correct circuit passes through a predictable state but continues to operate thereafter. Signals are expressed by constants, simple terms, and initialized signals:

$$\begin{array}{c} \boxed{v} = \langle v, \boxed{v} \rangle \\ \boxed{f}(S_1 \ldots S_n) = \langle f(S_1^{\bullet} \ldots S_n^{\bullet}),\ \boxed{f}(S_1^{\dagger} \ldots S_n^{\dagger}) \rangle \\ v\,!\,S = \langle v, S \rangle \end{array}$$

where $f : V^n \to V$ is a strict operation on the ground domain; and if $S = \langle v_0, v_1, v_2, \ldots \rangle$, then $S^\bullet = v_0$ and $S^\dagger = \langle v_1, v_2, \ldots \rangle$.

The boxes distinguish the behavioral interpretation of the underlying symbols from their discrete interpretation in specifications. Although the discrimination is sometimes helpful, it is unnecessary when reasoning about terms. The elementary algebra for terms, based on composition, combination, construction, and substitution, commutes with the signal-interpretation. For example,

$$\boxed{f \circ g}(S) \equiv \boxed{f}(\,\boxed{g}(S)), \quad \text{and} \quad \boxed{\lambda(u,v).f(u,g(v,u))}(X,Y) \equiv \boxed{f}(X,\ \boxed{g}(Y,X)).$$

This transparency is the basis of the argument for using a functional calculus in digital engineering.

A *circuit description* is a set of mutual signal-defining equations, for example,

$$\begin{aligned} X &= n\ !\ \boxed{decrement}(X) \\ Y &= 1\ !\ \boxed{multiply}(X,Y) \\ P &= \quad \boxed{zero?}(X) \end{aligned}$$

This recursive system abbreviates the recurrence relation

$$\begin{array}{ccc} X_0 = n & Y_0 = 1 & \\ X_{k+1} = X_k - 1 & Y_{k+1} = X_k \times Y_k & P_k = (X_k = 0) \end{array}$$

The model says nothing about time, but if it is assumed that values with the same index are coincidental, then the example describes a clocked sequential circuit, or *synchronous system* [4]. Initialized signals denote clocked storage elements; other terms are combinational. All feed-back loops pass through a storage element. There is an obvious schematic correspondence; signal identifiers name paths whose connectivity is prescribed by the equations:

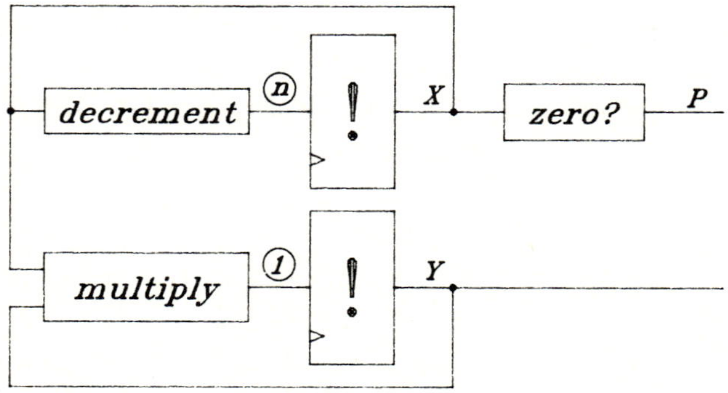

A Synthesis Technique

The circuit description

$$
\begin{aligned}
X &= n \,!\, \boxed{decrement}(X) \\
Y &= 1 \,!\, \boxed{multiply}(X, Y) \\
P &= \boxed{zero?}(X)
\end{aligned}
$$

is a *transcription* of the specification

$$
\begin{aligned}
&F(n, 1) \\
&\quad where \\
&\quad F(x, y) \;\Leftarrow\; zero?(x) \rightarrow y,\; F(decrement(x), multiply(x, y)).
\end{aligned}
$$

The approach to circuit synthesis is based on the following valid generalization [9]:

If p, r, and t_i are simple terms then

$$F(x_1, \ldots, x_n) \;\Leftarrow\; p \rightarrow r,\; F(t_1, \ldots, t_n) \qquad (*)$$

is realized by the circuit description

$$
\begin{aligned}
X_1 &= v_1 \,!\, t_1 \\
&\;\vdots \\
X_n &= v_n \,!\, t_n \\
RDY &= p \\
ANS &= r
\end{aligned}
$$

That is, signal ANS contains $F(v_1, \ldots, v_n)$ as soon as signal RDY contains true.

Thus, the example computes $n!$ if its storage elements are initialized to n and 1. This can be proved directly, but it also follows from the generalization. Specification (*) is essentially a schematic for a synchronous system that computes the function (*). The approach to synthesis, then, is to derive a correct instance of (*) from an arbitrary initial specification, using transformations.

If the initial specification is iterative, an equivalent instance of (*) can be mechanically derived. That construction corresponds to the structured digital design technique of decomposing a circuit into an architecture governed by a controller. In digital design, control is usually described by a finite-state machine or a flowchart. Control hardware is systematically derived from the control description [4, 15]. Since iterative schema characterize "flowchartability", writing an iterative specification is equivalent to conceiving a sequential-control algorithm.

Non-iterative specifications represent a higher level of specification that is not available in the conventional methodologies. Their direct translation to circuits is not always possible; non-linear recursions must be removed or implemented. For both tasks, systematic formal techniques exist from research in program transformation, program synthesis, denotational semantics, and comparative schemata.

An Example

This section develops a small example of circuit-description synthesis and demonstrates various uses of functional abstraction in the derivation of a realization. An iterative version of an initial specification is obtained by implementing recursion with a stack. Since the initial specification is linear, a stack is not strictly necessary. However, its incorporation provides a context for a discussion of the treatment of complex data types. The initial steps of the derivation follow a standard pattern and are presented with little justification; more complex exercises can be found in [8] and [9], which cite related research. The interested reader might begin with [13] and [2] and find mathematical foundations in [7] and [11].

- Let us take as a specification the general linear recursion scheme

$$F(x) \quad \Leftarrow \quad p(x) \rightarrow f(x),\ g(x, F(h(x))).$$

The variables p, f, g, and h stand for arbitrary combinations of strict operations on a fixed but unspecified ground domain. The example is quite general: any linear recursion equation can be expressed in this form. (The *factorial* specification shown earlier is an instance of this scheme.)

- Introduce a continuation—an abstraction for control—to specify an order for the computation.

$$F(x, [\lambda v.v]) \ \textit{where}$$
$$F(x, \gamma) \quad \Leftarrow \quad p(x) \rightarrow \gamma f(x),\ F(h(x),\ [\lambda v.\gamma g(x, v)]).$$

Landin was the first to use functional abstraction to describe control in this way [5]. In denotational semantics, this is a standard step in the development of an operational semantics for a programming language [11].

- **Choose a stack to represent the continuation and define a function to interpret that representation.**

$$\begin{array}{l} F(x, empty) \ \textit{where} \\ \quad F(x,c) \ \Leftarrow \ p(x) \rightarrow G(f(x),c), \ F(h(x), \ push(x,c)). \\ \quad G(v,c) \ \Leftarrow \ empty?(c) \rightarrow v, \ G(g(top(c),v), \ pop(c)). \end{array}$$

(For the *factorial* instance of this specification, an accumulator is a better representation choice, giving rise to the iterative version of *Fac* shown earlier [13].)

- **This version of *F* is iterative. Add a token encoding which of the functions *F* and *G* is "in control".**

$$\begin{array}{l} F(\mathtt{F}, x, empty) \\ \quad \textit{where } F(s,x,c) \ \Leftarrow \\ \qquad (s = \mathtt{F}) \rightarrow \\ \qquad\quad [p(x) \rightarrow F(\mathtt{G}, f(x), c), \ F(\mathtt{F}, h(x), push(x,c))] \\ \qquad (s = \mathtt{G}) \rightarrow \\ \qquad\quad [empty?(c) \rightarrow v, \ F(\mathtt{G}, g(top(c),x), \ pop(c))]. \end{array}$$

- **An instance of the desired scheme is obtained by distributing the conditional across the calls to *F*. The general distributive law is**

$$p \rightarrow f(x), \ g(y) \quad \equiv \quad (p \rightarrow f, g)(p \rightarrow x, y)$$

First, define a combinator ("macro") for the selection operation:

$$sel \stackrel{\text{def}}{=} \lambda(q, r, v_0, v_1, v_2).\ q \rightarrow [r \rightarrow v_0, v_1], v_3$$

With some rearrangement and identification of terms, distributing *sel* yields

$$\begin{array}{l} F(s,x,c) \ \Leftarrow \\ \quad \textit{let } q = equal?(s, \mathtt{F}) \\ \qquad\ \ r = p(x) \\ \qquad\ \ t = top(c) \\ \qquad\ \ e = empty?(c) \\ \quad \textit{in} \\ \quad and(not(q), e) \rightarrow x, F(\ sel(q, r, \mathtt{G}, \mathtt{F}, \mathtt{G}), \\ \qquad\qquad\qquad\qquad\qquad sel(q, r, f(x), h(x), g(t,x)), \\ \qquad\qquad\qquad\qquad\qquad sel(q, r, c, push(x,c), pop(c))\ ). \end{array}$$

- **This version of F is in the characteristic form and is realized by**

$$\begin{aligned}
F(s^0, x^0, c^0) \;&\Leftarrow\; (X,\ RDY)\ \textit{where} \\
S &= s^0 \ !\ \boxed{sel}(Q, R, \boxed{\mathtt{G}}, \boxed{\mathtt{F}}, \boxed{\mathtt{G}}) \\
X &= x^0 \ !\ \boxed{sel}(Q, R, \boxed{f}(X), \boxed{h}(X), \boxed{g}(T, X)) \\
Q &= \quad \boxed{equal?}(s, \boxed{\mathtt{F}}) \\
R &= \quad \boxed{p}(X) \\
RDY &= \quad \boxed{and}(\boxed{not}(Q), E) \\
T &= \quad \boxed{top}(C) \\
E &= \quad \boxed{empty?}(C) \\
C &= c^0 \ !\ \boxed{sel}(Q, R, C, \boxed{push}(X, C), \boxed{pop}(C))
\end{aligned}$$

Abstract Components

Signal C in the derived circuit description ranges over stacks and, therefore, could hardly be regarded as a true data path. The factorization below yields a circuit over more concrete signals by introducing an information-hiding component analogous to an abstract data type. We continue the example, now concentrating on the complex signals

$$\begin{aligned}
T &= \quad \boxed{top}(C) \\
E &= \quad \boxed{empty?}(C) \\
C &= c^0 \ !\ \boxed{sel}(Q, R, C, \boxed{push}(X, C), \boxed{pop}(C))
\end{aligned}$$

The goal is to replace the signal C with an *abstract component* which manages the stack. At the same time, the surrounding circuit should retain the selector that determines what is done to the stack. The key step is to introduce a "signal" that ranges over operations and is applied to the appropriate operands.

$$\begin{aligned}
C &\simeq c^0 \ !\ \boxed{I}(X, C) \\
\boxed{I} &\simeq \quad \boxed{sel}(Q, R, \boxed{\lambda(v,c).c}, \boxed{\lambda(v,c).push(v,c)}, \boxed{\lambda(v,c).pop(c)})
\end{aligned}$$

This gets the selection out of C's definition, but signals don't range over operations. $\boxed{I}$ is really a signal of instructions. Define an interpretation for instruction-representations.

$$interpret \stackrel{\text{def}}{=} \lambda(i,x,c)\,.\,(i = \text{STET}) \rightarrow c,\quad (i = \text{PUSH}) \rightarrow push(x,c),\quad (i = \text{POP}) \rightarrow pop(c).$$

Instruction-signal I becomes $\boxed{sel}(Q, \boxed{\text{STET}}, \boxed{\text{PUSH}}, \boxed{\text{POP}})$. Now define the abstract component *STACK* to be

$$STACK\ (c^0, I, X) \Leftarrow (\boxed{empty?}(C), \boxed{top}(C))$$
$$\textit{where}\ C = c^0\ !\ \boxed{interpret}(I, X, C)$$

With the incorporation of *STACK*, the circuit description is now in terms of concrete signals, schematically depicted below.

$$F(s^0, x^0, c^0) \Leftarrow (X,\ RDY)\ \textit{where}$$
$$S = s^0\ !\ \boxed{sel}(Q, R, \boxed{\text{G}}, \boxed{\text{F}}, \boxed{\text{G}})$$
$$X = x^0\ !\ \boxed{sel}(Q, R, \boxed{f}(X), \boxed{h}(X), \boxed{g}(T, X))$$
$$Q = \boxed{equal?}(s, \boxed{\text{F}})$$
$$R = \boxed{p}(X)$$
$$RDY = \boxed{and}(\boxed{not}(Q), E)$$
$$I = \boxed{sel}(Q, R, \boxed{\text{STET}}, \boxed{\text{PUSH}}, \boxed{\text{POP}})$$
$$(E, T) = STACK(c^0, I, X)$$

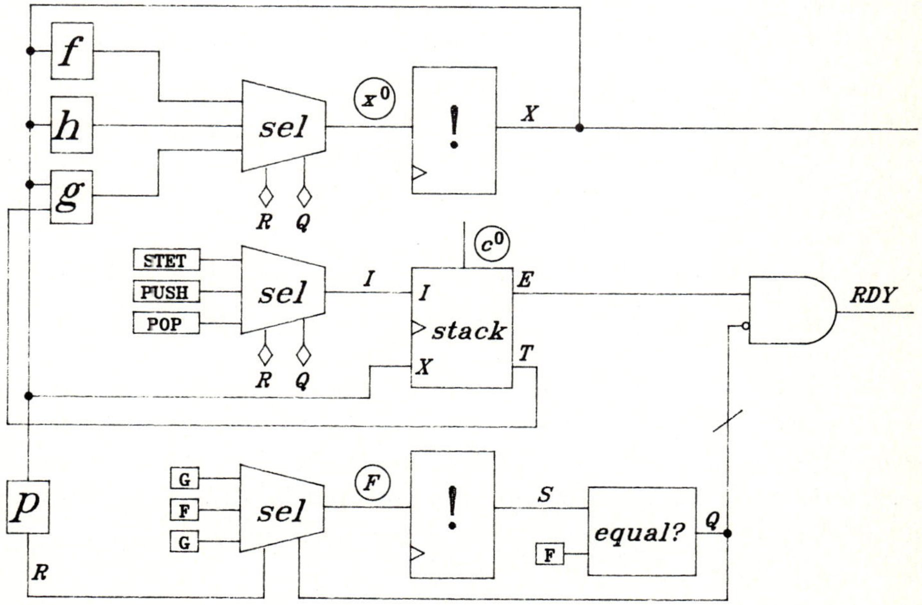

Remarks

This synthesis technique does not account for all circuits, nor does it generate complete circuit descriptions. Its utility in the description of digital filters and systolic circuits, for example, is debatable because control is, in the first case, usually iterative, and, in the second case, usually distributed. The circuitry needed to interface, power, and clock the implementation is not derived. The technique is better suited to problems in which the algorithmic complexity is significant, such as the implementation of a programming language processor—this approach was instigated, in part, by results in compiler/machine synthesis (e.g. [14]).

The strong connection to conventional methodology is best seen in comparison with ASM design techniques, where flowchart notation is used to address control [15]. Under the assumption of synchrony, synthesis by transformation ensures the construction of a clocked-sequential system; it is impossible to introduce combinational loops. In this respect, the functional approach maintains a stronger (syntactic) relationship between controlling and controlled architecture. The ASM method properly views the control algorithm as the dominant design aspect. The controlled architecture is, in principle, implied by the controller. Similarly, the transformation technique centers on the derivation of a sequential-control specification. The connectivity of controlled devices is a by-product of the derivation.

In the example, an external, communicative description for the abstract component *STACK* is also derived. This is one way to develop specifications that, unlike the example, have signals as parameters. *STACK*'s *implementation*, a description of its internal signals, may require a different catalog of techniques. Such objects are usually specified using algebraic methods, whose extension to behaviors is being studied.

The *STACK* factorization points out inadequacies in the simple model of behavior. Physical signals *are* sequences of concrete values, but a higher-order interpretation is useful for decomposition and refinement. Define a *component* to be a signal-of-operations and extend the definition of application. Informally,

$$\boxed{f} = \langle f, \boxed{f} \rangle$$

$$C(X_1 \ldots X_n) = \langle C^{\bullet}(X_1^{\bullet} \ldots X_n^{\bullet}),\ C^{\dagger}(X_1^{\dagger} \ldots X_n^{\dagger}) \rangle$$

This model supports the steps leading to the definition of *STACK* in the example, but also raises problems in the treatment of multiple-valued or partial components and in the assurance of a uniform signature. In practice, these problems are avoided by constraining the way in which components are introduced, just as combinational feed-back is avoided by the transformation process. The component $\boxed{I}$ was composed of strict functions on a normalized parameter.

A derived circuit description is still subject to refinement. Transformations on signals, such as those developed by Sheeran [10], seem appropriate at this level, as does reasoning based on interval logic. The theorem giving a characteristic specification for synchronous systems establishes the temporal assertion, "*ANS is* $F(v_1, \ldots, v_n)$ *as soon as RDY is true.*" An algebra for local refinement should maintain this correctness property. It is reasonable to assume that some abstract-component *implementations* come with their own clocks. *STACK* was incorporated as a synchronous sub-circuit. To be useful in practice, the technique must be extended to cope with the implied communication protocols.

Circuit descriptions are directly interpretable as programs, with signals being represented as recursively defined lists. The resulting logical simulation is a data structure that can be manipulated with an editor. Figure 1, which shows an editor for the *factorial*-circuit written in Daisy, illustrates how little need be involved in simulating logical behavior. This kind of experimentation can reveal unforeseen properties of behavior. In the example, $n!$ appears one cycle before *RDY* is asserted, suggesting the addition of a *zero?*-test to the output of *decrement*. Signal *Y* eventually becomes stable, a fact which may be useful in designing an interface to the circuit.

Acknowledgments

This material is based on work supported by the National Science Foundation under grant number DCR84-05241. John O'Donnell contributed key insights, which I hope are adequately reflected here. I am indebted to Franklin Prosser and David Winkel for their guidance and encouragement.

References

[1] Abelson, H. and Sussman, G. J., *Structure and Interpretation of Computer Programs* (The MIT Press and McGraw-Hill, Cambridge, 1985).

[2] Darlington, J. and Burstall, R. M., A system which automatically improves programs, *Acta Informatica,* **6** (1976) 41–60.

[3] Henderson, P., *Functional Programming: Application and Implementation* (Prentice-Hall, Englewood Cliffs, 1980).

[4] Hill, F. J. and Peterson G. R., *Introduction to Switching Theory and Logical Design* (Third Ed.) (John Wiley&Sons, New York, 1981).

[5] P. J. Landin, An abstract machine for designers of computing languages, *Proc. IFIP Congress 65,* Vol. 2, (Spartan Books, Washington; MacMillan, London, 1966)

[6] McCarthy, J., A basis for a mathematical theory of computation, in: P. Braffort and D. Hirschberg (eds.), *Computer Programming and Formal Systems* (North-Holland, Amsterdam, 1963).

[7] Manna, Z., *Mathematical Theory of Computation,* (McGraw-Hill, New York, 1974).

[9] Johnson, S. D., Applicative Programming and Digital Design, *Eleventh Annual ACM SIGACT-SIGPLAN Symposium on Principles of Programming Languages* (1984).

[9] Johnson, S. D., *Synthesis of Digital Designs from Recursion Equations, The ACM Distinguished Dissertation Series,* (The MIT Press, Cambridge, 1984).

[10] Sheeran M., Designing regular array architectures using higher order functions, in: Jouannaud, J-P (ed.), *Functional Programming Languages and Computer Architecture* (Springer [*LNCS* No. 201], Berlin, 1985).

[11] Stoy, J. E., *Denotational Semantics: The Scott-Strachey Approach to Programming Language Theory* (The MIT Press, Cambridge, 1977).

[13] Wand M., Continuation based program transformation strategies, *J. ACM* **27**(1) (1980) 164–180.

[14] Wand, M., Deriving target code as a representation of continuation semantics, *ACM Trans. on Programming Languages and Systems* **4** (1982) 496–517.

[15] Winkel D.E., and Prosser F.P., *The Art of Digital Design* (Prentice-Hall, Englewood Cliffs, 1980).

```
FACckt:n <= <X Y RDY>
  where
    X = < n ! DECREMENT:<X> >
    Y = < 1 ! MULTIPLY:<X Y>>
  RDY =       ZERO?:<X>

Query:[ [   Q     ! Qs]
        [[X Y R] ! Ss] ] <=

  let Next = Query:<Qs Ss>
      Now  = Query:<Qs <<X Y R> ! Ss> >
   in
   if:< same?:<Q "+">  Next
        same?:<Q "$">  <"exit">
        same?:<Q "?">  <"X=" X
                         "Y=" Y
                         "R=" R ! Now>
        same?:<Q "X">  <X ! Now>
        same?:<Q "Y">  <Y ! Now>
        same?:<Q "R">  <R ! Now>
                       Now
     >
```

(a)

```
& Query:< console:"FAC& "
              Xps:FACckt:10    >

FAC& ?
[X= 10 Y= 1 R= []
FAC& +++

FAC& RYX
 [] 720 7
FAC& +Y+Y+Y
 5040 30240 151200
FAC& ?
 X= 4 Y= 151200 R= []
FAC& +++?
 X= 1 Y= 3628800 R= []
FAC& +?
 X= 0 Y= 3628800 R= T
FAC& +++

FAC& R
 []
FAC& ?
 X= -3 Y= 0 R= []
FAC& +?+?+?
 X= -4 Y= 0 R= []
 X= -5 Y= 0 R= []
 X= -6 Y= 0 R= []
FAC& $
 exit]
&
```

(b)

Figure 1

Daisy Experiment with the Factorial Circuit Description

The colon denotes application; **< ⋆...⋆ ! ⋆ >** *builds lists; '&' is Daisy's prompt. The primitive* **console**, *whose argument is a prompt, produces a character-stream from the keyboard;* **same?** *compares values;* **if** *is like Lisp's* **COND**.

DECREMENT, **MULTIPLY**, *and* **ZERO?** *are arithmetic operations extended to "streams" (lists), and* **Xps** *transposes a list-of-streams to a stream-of-lists. Each is trivial to define owing to Daisy's interpretation of list-application [9].*

FACckt *models the factorial-circuit.* **Query** *is a stream-oriented interpreter for questions about* **FACckt**'*s behavior. The source (a) has been modified; Daisy's syntax is slightly more crude. The transcript (b) is verbatim.*

Formal Aspects of VLSI Design
G.J. Milne and P.A. Subrahmanyam (editors)

The Algebraic Basis of an Expert System for VLSI Design

P.A.Subrahmanyam
AT&T Bell Laboratories Research

ABSTRACT

An algebraic basis for a transformation methodology that supports the design of provably correct integrated software-hardware systems is developed. The paper addresses primarily the higher levels of system design. The transformations proceed from an abstract behavioral description of the problem and result in a detailed architectural description at the chip level; such descriptions are then used as inputs to lower level silicon compilation tools for producing chip fabrication masks. The framework developed provides the ability to reason about the syntax and semantics of computational networks and their interrelationships to performance measures and network geometry; in addition, it supports a gradual introduction of system timing and synchronization information as a design evolves.

1. Introduction

A number of VLSI design tools (mainly addressing lower level design tasks) have recently been spawned to reduce the complexity of the VLSI design process. It has also been cogently argued that there is an urgent need for such tools to be less ad hoc if there is to be an improvement in the reliability of the resulting designs e.g., [5] ; unfortunately, however, few design aids fall into this class. This paper discusses a formal algebraic framework underlying an expert system for VLSI design[20]. Abstract behavioral specifications used as input are transformed into low level structural descriptions which are then converted into a physical (CMOS) layout. The objective of the system is to facilitate cost effective special purpose VLSI design while moving closer to the performance achieved by human experts. The major motivation for using a formal framework is to guarantee logical correctness of the designs generated. Additionally, the use of a cohesive formal basis that supports a broad spectrum of abstractions involved in the VLSI design process, from axiomatic behavioral specifications to switch-level MOS networks, is aimed at reducing the overhead of developing a family of support tools. Adopting an expert system paradigm provides flexibility in the system structure (in that incremental modification is facilitated) and enables improved feedback --both for programs and humans-- during the design process e.g, if some performance criteria are not met, it is possible to identify plausible causes for this failure and suggest remedies. It is important to note that by adopting an expert system paradigm we do not preclude algorithmic components; in fact, it is typically expected that some of the heuristic components will be replaced by algorithmic ones as time progresses.

1.1 Outline of the Paper

The next section provides a flavor of some of the issues addressed in the rest of this paper by means of an example. Section [3] reviews the major components of a system description and introduces the description primitives that we use. Section [4] introduces transformation schemas that enable a transformation of functional descriptions into computational networks. Section [5] illustrates the application of both the transformation schemas and geometric rules of inference in the generation of architectural descriptions. Section [6] sketches the manner in which system timing details and synchronization are addressed. Section [7] outlines the components of the expert system that convert structural descriptions into physical layout and the current implementation status. Section [8] elaborates upon the role of performance criteria in the design process. Finally, section [9] summarizes the results of the paper, and delineates ongoing research.

2. An Example

To introduce some of the concepts involved, we consider a specification of a chip that sorts the elements in a list. The chip, which we call SortChip, accepts (i.e., inputs) a list l at the port labelled *InputList* (denoted InputList$\leftarrow$ l), and outputs a sorted version of the list at the port labelled *OutputList* (denoted

OutputList→ Sort(l)). This behavior may be described by a "behavior expression" as follows:

SortChip = InputList← l ~> OutputList→ Sort(l)

where l is an instance of a list and Sort is a function defined on lists. The symbol "~>" may be read as "followed by", and indicates the fact that input is first accepted at the port InputList, and that this is then followed by the output of the sorted list at the port OutputList. If, as we would normally desire, the chip repeats this cycle of behavior, i.e., is now ready to accept a new list to be sorted, this can be expressed by the recursive behavior expression

SortChip = InputList← l ~> OutputList→ Sort(l) ~> SortChip

More formally, objects of the type List (which in turn consist of objects of some generic type Element) can be defined axiomatically as usual (a detailed specification is given in Appendix I). In particular, the definition of the function Sort may take the following form:

Sort(l) = if l=NewList then NewList else Add(Sort(Delete(l,Max(l))), Max(l))

We will illustrate in this paper how, as a result of applying transformations, the function Sort as defined above can be implemented by a triangular or a linear network of processing elements as shown in Figure 1. We can algebraically reason about the "shape" of such networks. For example, it can be shown that the first network has a shape that is "triangular", time complexity that is linear in N (the number of elements to be sorted) and an area complexity that is O(N^2). The second network has area O(N) area complexity, throughput N+1, and has a reduced pinout requirement due to serial input of data.

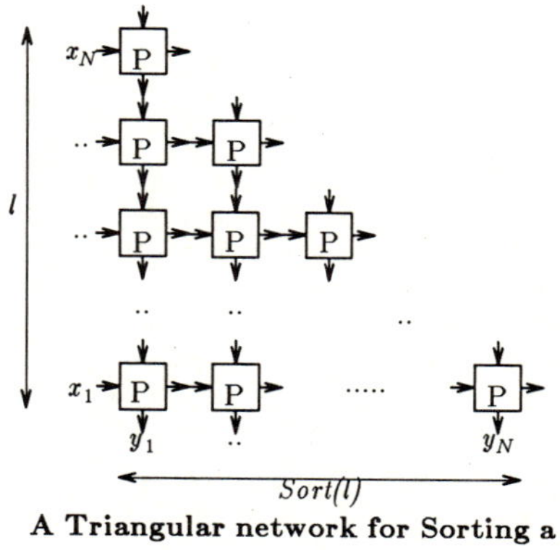

A Triangular network for Sorting a List

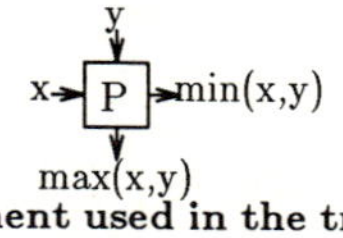

The processing element used in the triangular Sorter

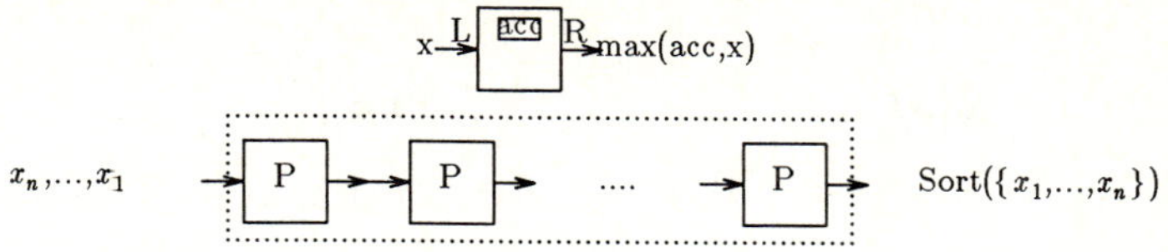

Figure 1. Examples of Linear and Triangular Network Implementations of the function Sort

A more detailed discussion of the above networks and alternative structures is contained in Sections [5,7].

When an architecture has been detailed to a level where the primitives have a known structure, the structure is first converted to a transistor connectivity list and then to a symbolic cell by a leaf cell layout program called Topologizer [17]. If desired, some technology dependent optimizations can be applied just prior to this stage; it is possible to use an algebraic model even at this level of abstraction e.g., [16]. In the case that the complexity of the leaf cell exceeds the capabilities of Topologizer, an intervening set of layout and floorplanning experts is used to iteratively converge towards a satisfactory layout [15]. The design of the set of interacting experts permits information in the form of constraints to be exchanged in such a way that the overall design methodology is top-down, *with bottom-up "suggestions" being accommodated;* we believe that this is an important characteristic of human design that is captured by this paradigm. In summary, an input behavior (e.g., the behavior expression for the Sort chip above) is converted into mask descriptions for circuit fabrication. We now proceed to summarize the salient features of the underlying formalism, preparatory to discussing algebraic transformations of behavior expressions into structural descriptions.

3. The Basic Model

3.1 System Components

3.1.1 Functionality. Viewed *externally*, a system (whether implemented in hardware or software or a combination) provides a set of functions.[1] Each function provided by the system can be viewed as a map from an input domain to an output domain F: D → D', for example Sort: List → List, Multiply: Matrix, Matrix → Matrix etc.

In an actual implementation, the abstract data values in the input and output domains have a specific representation in terms of the concrete data values in the input and output data streams that are input and output at the I/O ports of a system. For example, the elements that make up a list may be input either in parallel or as a sequential stream, the bits that represent an integer value may be input either serially or in parallel, the elements of a matrix may be input in row order, or in column order, or in diagonal order (as its bands) etc. This relation between the values in the input/output domains of a function and the values in the input/ output domains of a network that implements this function is embodied by the following two functions

(Input-Representation-Function) I: Abstract Input Domain → Input Data Stream
(Output-Representation-Function) O: Output Data Streams → Abstract Output Domain

In the next section, we will discuss some I/O representation functions in detail.

Internally, the system provides its functionality by virtue of the finer grain structure it possesses. The

1. We use the word function somewhat loosely to include nondeterministic operators, although most of the theoretical development is this paper focuses only on deterministic functions.

internal structure of a system typically consists of:

- the constituent processing elements
- the interconnections between these processing elements

The constituent processing elements in turn provide similar functions, f_i: $D_i \rightarrow D_i$', where the D_i's and the D'_i's are related to D and D' through the representation functions I and O. In essence, the $\{f_i\}$'s combine to implement F.

A network that implements a function (or a set of functions) will be described using the primitives that we will outline below. Let us first briefly summarize what it means to say that a network implements a function. We do this by means of a simple example; further examples will follow later in this section. Consider a function f: T → T', where f(t)=t', and let I and O be the input and output representation functions used for the input and output of T and T'. Consider also a network N with two ports P and P' that are used for input and output of t and t' respectively. Intuitively, this means that to compute f(t), we input the representation of t i.e., I(t) to the network port P; we will eventually obtain an output (denoted x) from the port P' which is in fact the representation of the desired result, i.e., the actual result is obtained by applying the "decoding" function, or the output representation function O to the output value x at the port P'. Formally, the network N implements f if

$$(P \rightarrow I(t) \mid N) = P' \rightarrow x, \text{ where } O(x)=f(t).$$

We note that it is possible for the same network to implement more than just one function. In such a case, the inputs and/or outputs associated with these functions are either diferent in the times they are manifest at the network ports and/or use differing (not necessarily disjoint) sets of ports.

We next outline the primitives used for describing networks, and then discuss transformations for map functional specifications into networks.

3.1.2 Description Primitives and Semantics The basic description primitives we adopt here are a variant of those in a Calculus for Communicating Systems (CCS) introduced by Milner [11].[2] The primitive operations used permit

- *Sequential computations* e.g., function application, conditionals, etc. The functions themselves may be defined equationally.
- *Sequential Composition* (denoted X ~> Y), which denotes a system in which process X is *followed by* Y.
- *Input/Output* at ports (denoted by InputPort← value and OutputPort→ expression respectively). When a value is output at an output port, it is communicated to all of the input ports connected to it.
- *Parallel Composition* (par(X,Y), denoted X||Y). X||Y denotes a system in which both X and Y are scheduled in parallel, and whose actions may be interleaved in arbitrary fashion.

  The first action in X||Y may either be from X or Y, or may involve simultaneous actions in X and Y. In order to express this, we introduce two auxiliary composition operations.

  a. *Sequential-Parallel Composition.* seqpar(X,Y) (synonymously, X<|Y) denotes a concurrent system in which the first action is chosen from X, but following which the remaining actions in X and those in Y may be arbitrarily interleaved (i.e., as in par).

  b. *Composition.* compose(X,Y) enforces a synchronization between the processes X and Y. This synchronization may be accomplished either by

     i. a communication of values between X and Y (via their input/output ports), or by

2. Our specific choice, while convenient, is by no means essential to the discussion in the rest of this paper.

ii. a "lock step" (concurrent) execution of actions in X and Y (for example, one that is controlled by a global clock).

We will denote communication synchronization by X|Y and concommitant actions by [X,Y] for the intuition we hope these symbols convey.

- *Abstraction* (Encapsulation). If A is a set of actions, X\A denotes the system X with the actions in the set A being precluded. Intuitively, abstraction enforces certain restrictions on the mode of communication between modules *e.g.*, by "hiding" the communication ports that enable actions in A.
- *Iteration/Recursion*. Several interesting processes are "iterative" or recursive in their behavior. Their description may be presented via a recursive expression. While intuitive examples will be provided shortly, we defer to [11] for a formal exposition.
- *Nondeterminism* or *choice* action or *selection* (denoted by +). X + Y means that either the behavior X or Y can be nondeterministically chosen.[3]

Composition of behaviors (denoted by B1 | B2), involves a "hooking up" of I/O ports (denoted by P1 — P2, when the two ports P1 and P2 are connected).

A detailed description of variants of these primitives and associated semantics may be found in [11] , and we will not elaborate upon them any further here. An equational specification of these operations is summarized in Appendix [Composition Primitives] (cf. [22]).

Auxiliary operations that are useful in system description include:

1. Relabeling (denoted P1/Q1 when the port named P1 is renamed to be Q1).
2. Hiding of port names (denoted B \ {P,Q,...} when the ports P, Q, ... used in the definition of the behavior B are to be hidden from the external world, i.e., serve purely as internal ports of a subsystem).

We will adopt the convention that if an input and output port have the same name, then they are connected to each other (such ports are called "complementary ports" [11]).

For instance, if an input port P1 is connected to an output port P2, and a value, say 5 is output at P2 and is accepted (i.e., input) at P1, then this communication is denoted by writing

$$P2\!\rightarrow 5 \mid (P1\!\leftarrow x \sim> P3\!\rightarrow x * x)\ [P1\text{—}P2] = \tau \sim> P3\!\rightarrow\ 5*5$$

where the symbol τ is used to denote the action of communication of the value 5 between the ports P1 and P2 (and the associated time elapsed).

In the examples that follow, we make use of a tupling operator $<....>$ that can enclose a set of input/output actions. Operationally, the semantics of the tupling operator can be informally thought of as allowing the set of actions in the tuple to potentially (but not necessarily) occur in parallel. More precisely, given the behavior expression $<action_1,\ action_2,\ ....action_n> \ \sim> \ B1 \ \sim> \ B2$ only those actions in the set $\{<action_1,\ action_2,\ ....action_n>\}$ that are required for B1 need occur (either in parallel or sequentially in arbitrary order); the remaining actions in the tuple are then carried through to precede the behavior B2, and so on. In other words, the tupling operator specifies a form of data driven computation. A detailed description of the semantics of the tupling operator as well as specialized methods for reasoning about regular computational networks are discussed in a companion report.

4. Obtaining Networks from Functional Specifications

3. We will not make explicit use of the nondeterminism operator in this paper.

4.1 Preliminaries

Given the definition of a function, we will now discuss transformations that yield a computational network that implements this function. We will later discuss how one can reason about both the functional and the geometric properties of these networks.

In this paper, we assume, without loss of generality, that functions are defined on some (set of) abstract data types e.g., integers, stacks, lists, trees, etc. Such function definitions are typically given by specifying the action of the function casewise on a set of prototypical instances of the type. These prototypical instances mirror the structure of the objects e.g., a list is either an empty list (denoted here by NewList), or is obtained by adding an element to a list (denoted by Add(l,x)).

For the examples in the rest of this section, we will assume that

1. NewT is a base constructor that serves to spawn new instances of the type T, with arity[4] NewT: X $\rightarrow$ T where X is a (possibly empty) list that does not contain T, e.g., NewStack: IntegerSize $\rightarrow$ Stack, NewList: $\rightarrow$ List, etc.
2. c is a constructor for instances of the type T, with arity c: T,X$\rightarrow$ T, where X may again be a list of types e.g., Push: Stack, Integer $\rightarrow$ Stack; Successor: Integer $\rightarrow$ Integer; Add: List, Element $\rightarrow$ List; MakeTree: Tree, Item, Tree $\rightarrow$ Tree etc.

In essence, these assumptions imply that all instances of type T can be generated by applications of NewT and c, and are of the form {NewT(x_1), c(NewT(x_1), x_2), c(c(NewT(x_1), x_2),x_3), ... | x_1,x_2, ... $\in$ X}. It is therefore sufficient for any function to be defined on prototypical instances of type T that are of the form NewT(x_1), and c(t,x). This class includes a number of the commonly used data types e.g., integers, arrays, lists, trees, etc.

We will further assume that a space metric M is defined on T as follows

M(NewT) = 0
M(c(t,x)) = M(t)+1

By saying that M is a space metric on T, we mean that M(t) defines a measure of the amount of space (area) needed to represent t. More precisely, M: T $\rightarrow$ R is a function from T to R (the real numbers) although it is often more appropriate to consider asymptotic approximations to the value of M that are suited to the level of abstraction under consideration. As an example, the function M defined above would imply that the space needed to represent a list is proportional to the number of elements in the list. It is important to note that M can be any arbitrarily complex function in general, although the definition we have chosen very often suffices at the level of abstraction we are primarily concerned with in this paper.

4.1.1 Serial and Parallel I/O Representations for the type T For the examples in this section that operate on lists, we will consider two representations of lists: as tuples of items in the list (usually used for parallel input) and as streams of items in the list (usually used for serial input). As we indicated earlier, lists embody the essential properties of the generic type T that we are considering in this section.

In the tuple representation, lists are represented as tuples of the items contained in the list (in the same order). The only difference is that we allow certain dummy items denoted "*" in the representation tuple which are ignored when interpreting a tuple as a list. Intuitively, this allows us to use a tuple of length n to represent lists with less than n "real" items (simply by dropping the dummy elements). This is convenient when it is necessary to interpret the outputs at a fixed number of ports to be a variable length item, *e.g.*, when we have a network that deletes items from a list.

4. The *arity* of a function is the list of the *names* of the input and output domains of a function.

More formally, if we have a type called Tuple, and "*" denotes the dummy item, the input and output representation functions I:List(X)→ Tuple(X) and O:Tuple(X)→ List(X) may be defined as follows:

I(NewList) = NewTuple(N)
I(Add(l,x)) = Append(I(l),x)

O(NewTuple(N)) = NewList
O(Append(t,x)) = if x="*" then O(t) else Add(O(t),x)

In several of the examples below, we will be considering lists represented by N-tuples, for a constant N. In such a case, the representation will be accurate only for lists of length at most N. Further, we will assume that the input representation function pads the N-tuple with *'s on the left, and represents the empty list by an N-tuples of *'s; we omit the formal definitions of these functions for brevity.

In the stream representation of a list, the items in the stream consist of the items in the list (in the same order). Again, we allow for dummy items in the streams, and denote them by "*". In addition, the end of a list needs to be indicated when using a stream representation in order to distinguish between the end of one list and the beginning of another; we will use the symbol "eos" to denote the end of a list. More formally, if the data type Stream (of items) has the base constructor EmptyStream and the constructor Append, the input and output representation functions I:List(X)→ Stream(X) and O:Stream(X)→ List(X) may be defined as follows:

I(NewList) = EmptyStream
I(Add(l,x)) = Append(I(l),x)

O(EmptyStream) = NewList
O(Append(s,x)) = if x="*" then O(s) else Add(O(s),x)

4.1.2 Some Notation The ports {P, P1, P2, ...} of the processing elements in M will be referred to either by their names P, P1, ... or, in case of possible ambiguity, by qualified port names e.g. or, if ambiguous, by their qualified names M.P, M.P1, ... etc.

When considering a network (module) M with an output port P, the actual value, say p, that is output at the port P will in fact be a representation (e.g., in terms of streams or tuples) of the values in the abstract output domain (e.g. lists). In proving the correctness of the schemas that follow, we will therefore have to apply the ouput representation function O to this value p. Rather than write "O(p)=x where M.P→ p", we will simplify notation by abbreviating this to O(M.P→)=x.

In a module M, if a port P inputs/outputs objects of type T, and we wish to explicitly denote this, we write M{...P:T...}. Further, if a port is a bus of width N that is used to represent a tuple of values of type T, we write M{...P[N]:T...}

We use the notation $\prod_{i=1}^{i=N} P_i$ to denote the composition $P_1 | P_2 | \cdots | P_N$ of N instances of a processing element P; the i-th instance of the processing element P in such a composition is denoted P_i.

It is often possible to succinctly express such compositions via appropriately defined higher-order functions. For example, we can define a function "pipeline" that accepts two sets of arguments as follows:

```
pipeline(m,n)(f,[i,o],p) = f(m) | f(m+1) | ... | f(n)
                           connect [f(1).i--f(2).o] .. [f(n-1).i--f(n).o]
                           hide {f(1).i, .... f(n).o}, where m≤n
```

or simply as

```
pipeline(m,n)(f) = f(m) | f(m+1) | ... | f(n), m≤n
```

In the next section, we will illustrate the use of such functions in specifying subnetworks of the sort network.

4.2 Examples of Transformation Schemas

Transformation schemas are used to provide implementations for (sub)classes of behavior expressions. Each transformation maps a behavior expression (which may initially be the input specification) into another behavior expression (which is ultimately a description of the final circuit or a part thereof). Each transformation is associated with a set of attributes that characterize the performance metrics of the network that results from applying this transformation. The attributes associated with a transformation are used to guide the overall transformation strategy.

The schemas used by the system can be categorized into two classes:

1. An extensible set of special purpose schemas for synthesizing implementations of specialized (sub)classes of function definitions (or behavior expressions).
2. A fixed set of general purpose schemas that are used as defaults in the case when the needed behavior is neither available as a primtive nor synthesizable using the extant list of special purpose schemas.

Some representative components of a typical schema are indicated below:

- input-behavior-expression (and associated function definitions),
- output-behavior-expression (the implementation),
- properties of the implementation,
- preconditions (that must be satisfied by the input, for application of this schema),
- postconditions (that hold after application of the schema).

A schema of the above form serves to provide an implementation for a function definition that matches a certain template — that of the input behavior expression. The output behavior expression part of the schema describes a network implementation that is the "result" of applying this schema; typically the network will depend upon the variables in the function definition that are instantiated when a schema matches an input behavior. The properties associated with a particular method of implementing the input behavior expression characterize various performance attributes of the schema, *e.g.*, whether it serves to decrease area, improve response time, etc. The preconditions (respectively postconditions) indicate the properties that must hold before (respectively after) the application of the schema.

We now enumerate some special purpose transformation schemas and indicate that nature of their proofs of correctness. General purpose schema are similar to compilation techniques, and we will not elaborate upon them here. We will focus here only on the input to the schema, and the module structure that is output. While our actual implementation uses a set of other schemas attributes, such as geometry, suggested floorplans, etc., we will not elaborate upon this any further in this paper. Examples of the application of these schemas will be detailed in the next section.

4.2.1 **Schema M**. Consider the function g: T→ X, defined as follows:

g(NewT) = k
g(c(t,x)) = if p(x,g(t)) then x else g(t)

where P has the functionality P: X,X→ Boolean. Then the behavior

GB{L:T,R:X} = L← t ~> R→ g(t) ~> GB{L:T,R:X}

can be implemented by the (primitive) network module GM{L:X, R:X} defined by

GM = GM1(k)
GM1(c) = L← x ~> if x=eos then R→ c ~> GM
elseif p(x,c) then GM1(x) else GM1(c)

using the I/O Representation functions:

I : T$\rightarrow$ Stream(X $\cup$ {eos})
O : Identity: X$\rightarrow$ X

where

I(t) = Append(Stream(t), eos)
Stream(NewT) = Nullstream
Stream(c(t,x)) = Append(Stream(t),x)
■

The behavior of the module GM is characterized by

$$L\rightarrow I(t) \mid GM = \ O(R\rightarrow) = g(t) \sim> GM.$$

i.e., GM exhibits the desired behavior GB. This essentially says that if the input t is encoded using I and presented at the port L of the module GM, then the output obtained at the port R of GM, when decoded using the function O, will be the desired result g(t). We will, for the moment, ignore other attributes of the schema such as area and time complexities, throughput, and the non functional ports introduced e.g., power and ground. It is important to note, however, that such attributes are integrated into the framework.

Proof of Schema M. To prove functional correctness of the schema, we need to show that the module GM exhibits the desirwed behavior, i.e., that L$\rightarrow$ I(t) | GM = O(R$\rightarrow$) = g(t) $\sim>$ GM.

The proof is by induction on the size N of the input, and therefore on the length of Stream(t).

Base Case. The base case occurs when N=0 i.e., t=NewT, and I(NewT)= <eos>.

L$\rightarrow$ eos | GM = R$\rightarrow$ k $\sim>$ GM

which yields O(R$\rightarrow$)= O(k) = g(NewT).

Induction Step. When N$\geq$1, t is of the form c(t',x), and Stream(c(t',x)) = Append(Stream(t'),x).

We assume as the induction hypothesis that if Size(t')=N-1, N$\geq$1, then

L$\rightarrow$ Stream(t') $\sim>$ L$\rightarrow$ eos | GM = R$\rightarrow$ g(t') $\sim>$ GM.

At the penultimate step of the above computation, we therefore have g(t') stored in GM1 i.e.,

L$\rightarrow$ Stream(t') | GM = GM1(g(t')).

We therefore have that

L$\rightarrow$ x|GM1(g(t')) = if p(g(t'),x) then GM1(x) else GM1(g(t'))

and consequently that

[L$\rightarrow$ Stream(t') $\sim>$L$\rightarrow$ x $\sim>$ L$\rightarrow$ eos] | GM = if p(x,g(t')) then R$\rightarrow$ x else R$\rightarrow$ g(t') $\sim>$ GM

which proves the correctness of the schema.■

Additional schema are detailed in the Appendix [Transformation-Schemas].

5. Applying Transformation Schemas: Some Examples

We now illustrate the application of the transformation schemas and the geometric shape inference rules in a few examples.

5.0.1 **Implementations for the function Max** Consider the definition of function Max on the type List(N)(Integer(B)) i.e., a list parameterized by integers of size B (cf. Appendix I) as reproduced below:

Max(NewList) = $-\infty$
Max(Add(l,x)) = if x > Max(l) then x else Max(l)

Given that we wish to implement this function, an applicable schema is one whose function syntax and semantics match the above. Schema M matches in this case, with the substitutions g=Max, T=List, X=Integer(B), NewT = NewList, c=Add, k=$-\infty$, P = > (greater-than). If a stream representation is used for the lists, Max may be implemented by a processing element which we will denote MaxPE that has ports {L:Integer(B), R:Integer(B)}, with the following semantics:

```
MaxPE = Max1(-∞)
Max1(c) = L←x ~> if x=eos then R→c ~> MaxPE
                       elseif x > c then Max1(x) else Max1(c).
InitialInput: L→Stream(l) ~> L→eos
Output: R→Max(l)
```

Properties of MaxPE. It is possible to deduce that MaxPE has area complexity 1, where the unit of area is the sum of the area required for storing an item and the processor area required for the computation of the conditional in Max1. The time complexity is O(M(l)), i.e., proportional to the number of items in the list, where the unit of time is the sum of the time required for input t port L, the time required for the computation of the conditional, and the time required for the output of the result at port R. Under the assumption that MaxPE has unit area, the definition of MaxPE satisfies the conditions for being a (primitive) linear array: it has two ports (L and R).

5.0.2 **Implementations for Delete** Consider the definition of the function Delete defined on a List (cf. Appendix I), as reproduced below:

Delete(NewList,x) = NewList
Delete(Add(l,x), x') = if x=x' then l else Add(Delete(l,x'),x)

This definition matches the definition of the function implemented by schema DT, with the substitutions F=Delete, p="=", T=List, NewT=NewList, c=Add, etc.

If we now assume that the processing elements used in the network implementations generated by these schemas are primitives i.e., of unit size, then the shape inference rules can be readily applied to infer their shapes.

The network Delete-L is obtained by using Schema DP; it has ports {L, R} and uses a stream representation of lists for input and output.

```
Delete-L = L←y ~> Delete-L1(y)
Delete-L1(y) = L←x ~> if x=eos then R→eos ~> Delete-L
      elseif x=y then Delete-L1(eos) else R→x ~> Delete-L1(y)
```

An alternative network can be obtained by applying Schema DT; we will denote this by Delete-R. The structure of Delete-R is given below:

```
Delete-R =  pipeline(1,N)(PM,connect([Left—Right]),hide({Left,Right}))
```

where PM has ports {Top, Bottom, Left, Right} and is defined by

```
PM = <Top←t, Right←r> if t=r then <Bottom→*, Left→*>
                               else <Bottom→t, Left→r>
```

If we assume that PM is a unit sized node, then it obeys the conditions for a primitive rectangular network [23]. Hence Delete-R can be shown to be a rectangular network by the inference rule for rectangular shapes. The output at the port Delete-R.Left is not needed in this case, and this port can therefore be hidden. Note that this has the effect of destroying the rectangular nature of Delete-R as far as further compositions are concerned, unless the dummy outputs * are retained.

5.0.3 **Implementation for Delete-Max** Consider now the deletion of the maximal element in a list i.e., the implementation of Delete(l,Max(l)). Using the semantics of Delete and Max, we obtain

Delete(NewList, Max(NewList)) = NewList
Delete(Add(l,x), Max(Add(l,x))) = if x = Max(Add(l,x)) then l else Add(Delete(l,Max(Add(l,x)),x)
= if x > Max(l) then l else Add(Delete(l,Max(l)),x)
[since Max(l,x) = if x>Max(l) then x else Max(l)]

This may therefore be rewritten as:

DeleteMax(NewList) = NewList
DeleteMax(Add(l,x)) = if x > Max(l) then l else Add(DeleteMax(l),x)

Using schemas that we will not elaborate upon here due to lack of space, DeleteMax can be implemented in several ways, two of which we illustrate below. If a stream representation of a list is used for input, then the functions DeleteMax and Max can be concomitantly implemented by the network P with ports {L,R} whose behavior is defined by

```
P = P1(-∞)
P1(c) = L←x if x=eos then R→eos ~> R→c P
                    elseif c=x then P1(x) else R→x ~> P1(c)
```

This network matches the definition of a linear array of size 1, with the left and right ports being L and R.

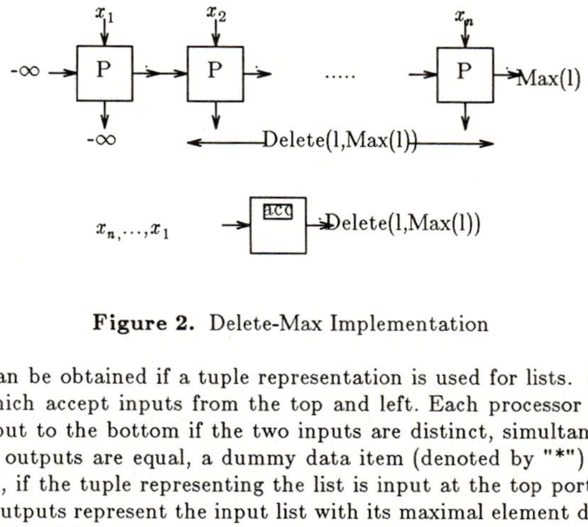

Figure 2. Delete-Max Implementation

An alternative network can be obtained if a tuple representation is used for lists. Intuitively, this yields an array of N processors, which accept inputs from the top and left. Each processor compares its two inputs, and transmits the top input to the bottom if the two inputs are distinct, simultaneously outputting its left input to the right. If the outputs are equal, a dummy data item (denoted by "*") is output at the bottom, and to the right. Overall, if the tuple representing the list is input at the top ports, and the left ports tied to -∞ , then the bottom outputs represent the input list with its maximal element deleted, and the rightmost element represents the maximal element in the list.

5.0.4 **Implementations of the Sort Function** Consider now the definition of the function given as below:

Sort(l) = Add(Sort(Delete(l,Max(l))),Max(l))

If we write Delete(l,Max(l)) as DeleteMax(l), this then yields:

Sort(l) = Add(Sort(DeleteMax(l)),Max(l))

But we have already derived implementations for Max(l) and DeleteMax(l)! Using them, and the fact that we can compose networks, we can now obtain several implementations for Sort that use the previously derived networks for Max and DeleteMax. We list three such networks below: the first has a triangular shape, and uses a tuple representation of lists for input and output; the second uses a linear representation of lists, and consists of cascading an appropriate number of PEs for Max and DeleteMax; the third consists of a single PE with a feedback loop. These are informally described below.

These networks are depicted pictorially in the figure below, and can be described as follows:

1. Network "Triangular Sort" is a triangular network, obtained by cascading N rectangular networks of size $1 \times i$, i varying from 1 to N (where N is the size of the list being sorted). Each of the cascaded rectangular networks computes the DeleteMax function on a list of decreasing size. The list is input as a tuple of its constituent elements. The sorted output is obtained at the edge of the triangle that is orthogonal to its input.
2. Network "Linear Sort" is a linear array N that accomplishes sorting by using local accumulators that store components of the final list. Again, each primitive component of this network consists of an element that implements the DeleteMax function on a lists of decreasing size. The original list is input as a stream at the left end of this array. The output can either be obtained in parallel after all of the list is input, or shifted out serially.
3. Network "Primitive PE Sort" is a single processing element that "sequentially" implements the sorting function, and is tantamount to the bubble sorting algorithm implemented using a single "internal register" and an I/O buffer of the size of the input! It is relevant to remark that we did not have these implementations of the sorting algorithm in mind when formulating these transformations, but that it transpired fortuitously.

5.0.5 A Comparator To indicate what happens when we descend to the next level of detail, let us consider the specification of the internals of one of the processing elements in the second sorting network enumerated above. The definition of the comparison operation defined on two integers of the same "bit width" can be defined as follows:

GreaterThan(0,0) = false
GreaterThan(AddHiBit(n1,b1),AddHiBit(n2,b2)) = if GreaterThanBit(b1,b2) then true
else if b1=b2 then GreaterThan(n1,n2) else false

Implementations obtained using this definition e.g., a ripple carry implementation for a comparator, can now be implemented using multiplexors and comparators and subjected to technology specific transformations (see Section [7]). Issues concerning strategy guidance are discussed in Section [8].

6. Timing Considerations

In this section, we will briefly mention some of the issues that concern system timing, i.e., the relative sequencing of operations and the behavior of signals over time. One of our major objectives in this context has been to avoid the premature introduction of explicit timing information during specification and/or synthesis. Typically, at the higher levels of system description, the major concern is system functionality, and the details of timing issues are best suppressed.[5] When dealing with system descriptions at a lower level, however, the temporal aspects assume greater importance and visibility. This perspective implies the need for a technique that supports gradual introduction of the relevant timing information as a design evolves. Further, it is important that the technique not impose any restrictions on the timing discipline adopted. More specifically, it should allow for synchronous and asynchronous (self-timed) circuits, local and global clocks, multiple clocks, and multiphase-clocks (such as those used in dynamic circuits).

5. There are, or course, exceptions to this rule, and consequently the ability to specify timing even at a higher level is provided.

6.0.1 Assumptions The main assumption made is that of causality: we assume that the output of a function (or module) can be computed (become available) only after the inputs it depends upon are available.[6] In addition to causality constraints introduced by functional dependencies, temporal constraints are imposed on a circuit's behavior by the finite nature of hardware resources (including wires and buses), and the manner in which they are shared by the modules in an implementation. We assume that a hardware implementation can be viewed as a state transition system, whether it be synchronous or asynchronous.

6.0.2 Notation For convenience, we will, in this section, use a "signal representation" when describing modules. That is, the ***port names*** will be implicit by their position, while the ***signal names*** at these ports will be explicitly named. For example, when we refer to a register with ports {In, Out} as Reg(*in*,*out*), it implies that *in* and *out* are the signals that are associated with the ports In and Out.

The temporal characteristics of a signal can be expressed using predicates that either explicitly depend on time, or, as is typical in temporal logic[10, 14] using predicates that implicitly depend on time.[7]

The temporal predicates that we use here are `always` (synonymously $\Box$) and `next` (synonymously $\bigcirc$). Since these predicates are standard in temporal logic literature e.g.,[10], we will not elaborate here on their formal semantics.

We now proceed to indicate how timing details may be introduced as the details of a design evolve.

6.1 Examples

6.1.1 Temporal Partial Orders. Consider a register with ports {In, Out} and the behavior:

$$\text{Reg} = \text{In} \leftarrow \text{x} \sim > \text{Out} \rightarrow \text{x} \sim > \text{Reg}$$

Its temporal behavior may be expressed as

$$\texttt{always}\ (\texttt{next}\ out = in)$$

Using an explicit interpretation of the "`next`" operator, this becomes

$$\text{for all t.}\ (out(\text{t}+1) = in(\text{t})).$$

Since we are dealing with a suitably discretized notion of time, these representations are essentially isomorphic.

A combinational circuit, e.g., a multiplexor mux with signals *select*, *in*1, *in*2, *out* can be characterized as follows:

$$\text{mux}(select,\ in1,\ in2,\ out) \equiv \texttt{always}\ (\text{if}\ select\ \text{then}\ out{=}in1\ \text{else}\ out{=}in2)$$

It should be noted that these temporal characteristics are independent of the initial behavior, except for the constraints implied by the causality principle.

6. Note, however, that while we always assume such a delay, no assumption is made as to whether it is "explicitly" acknowledged (as in sequential circuits) or "ignored" (as in combinational circuits).

7. An alternative way of expressing such designs is to make explicit a "clock" process, and include a clock port in the concerned modules[12]. We will not elaborate further on the details of such an approach in this paper.

6.1.2 Elaboration of Time Intervals; Timing Disciplines. The next level of timing detail in a design is introduced when the abstract "next" instance is interpreted (defined) as being the next time instance when some specific function (predicate) p:Time→ Boolean becomes true. In this context, it is useful to define the function NextTime as follows:

$$\text{NextTime(t1,t2)(p)} \equiv_{def} \text{t1} < \text{t2} \ \& \ \text{p(t1)} \ \& \ \text{for all t. t1} < \text{t} < \text{t2} \ \neg \text{p(t)}.$$

Intuitively, NextTime(t1,t2)(p) says that the signal p is true at time t1, and that the next time it becomes true is at t2.

If, for example, a clocked latch is used to implement a register, the predicate p may be interpreted as being true at the times "when the clock rises". If the function defining the clock behavior is *clk*:Time→ Boolean, then the first instance after t1 when the clock rises next is described by

$$\text{NextTime(t1,t2)(Rising}(\mathit{clk}))$$

where Rising(*clk*) is defined to be true at time t if the signal *clk* is true at time t and false just before.

In general, given a function s: Time→ LogicalSignalValues,[8] Rising(s) is defined to be true at time t if s is true at time t and false just before. The functionality of Rising is

$$\text{Rising: (Time} \rightarrow \text{Boolean)} \rightarrow \text{(Time} \rightarrow \text{Boolean).}$$

If a circuit is implemented asynchronously e.g., as in[8], and a request-acknowledge protocol is used, the predicate p would be defined to be true when the value of the *request* signal makes the desired transition.

Once the above timing details have been introduced, it is possible to amalgamate this into the register definition as follows:

$$\text{Reg}(\mathit{in},\mathit{out})(\mathit{clk}) \equiv \texttt{always} \text{ for all t1, t2 (NextTime(t1,t2)(Rising}(\mathit{clk})) \supset \mathit{out}\text{(t2)} = \mathit{in}\text{(t1))}$$

Note here that the parameter *clk* is now explicitly manifest in the definition of Reg, and the dependence on time of the signals in and *out* has now become more explicit; all of the signals *in*, *out*, and *clk* have the functionality: Time→ Boolean.

6.1.3 Constraints on Signal Behavior. The actual manner in which a module is implemented imposes further constraints on the behavior of signals over time, e.g., a signal might be required to be stable over some period of time. Typically, constraints are imposed upon the *input* signals (including the clock) of a module; given that the constraints on the input signals are met, the temporal characteristics of the output signals may be inferred from the details of the implementation.

Before illustrating the nature of such temporal constraints, we note that stability is normally defined as follows. Given a function f:Time→ V (for some arbitrary domain V)

$$\text{Stable(t1,t2)(f)} \equiv_{def} \text{for all t (t1} \leq \text{t} < \text{t2) f(t)} = \text{f(t1)}$$

embodies the fact f maintains a constant value from time t1 to just before t2.

The constraint (labelled in-Constraint) requiring the register input to be stable for some time time before the clock rises may now be expressed as follows:

8. The basic formalism we are using allows for multiple valued logic; however, we do not consider this any further in this paper.

$$\text{in-Constraint} \equiv_{def} \text{Rising}(clk)(\text{t1}) \;\&\; \text{Stable}(\text{t1-SetUpTime, t1})(in)$$

The implementation governs the value of SetUpTime, as also the values of certain parameters that influence how the clock should behave over time. Formally:

$$\text{clk-Constraint} \equiv_{def} (\text{for all t1,t2 [Rising}(clk)(\text{t1}) \;\&\; \text{NextTime(t1,t2)Rising}(clk) \supset \text{t2}-\text{t1} \geq \text{MinClkPeriod]}$$

Given these contraints on how the inputs *in* and *clk* behave over time, the behavior of the output over time may be inferred from e.g.,

out-Constraint ≡
 [in-Constraint &
 clk-Constraint &
 Rising(*clk*)(t1) &
 NextTime(t1,t2)(Rising(*clk*))
 ⊃ Stable(t1+StartTime, t2+FinishTime)(*out*) & *out*(t2)=*in*(t1)]

Note that both StartTime and FinishTime are defined by the details of the implementation.

At the lowest levels, the fabrication process and the detailed physical layout of a circuit influence the manner in which signals vary with time. A signal can then be considered to be a function from a suitably discretized time domain (Time) -- which we represent by the set of natural numbers -- to a set of signal voltage values V *i.e.*,

$$s: \text{Time} \rightarrow V$$

We further assume that the signal values can be further abstracted to a discrete domain of values *e.g.*, {O, 1, X} or {O, 1} in accordance with an appropriate circuit model. The levels of abstraction described earlier are then built up using this function.

6.2 Remarks

We have sketched a technique by which the details of system timing can be gradually introduced as a design progresses. At the highest level, no timing details need be explicit, and the only temporal constraints are causal. As a design evolves, and decisions relating to the implementation and timing disciplines are made, the temporal constraints on signals evolve alongwith. If, on the other hand, the external environment in which a system resides imposes any a priori temporal constraints, such constraints can be explicitly included in the initial specification, and accounted for in the synthesis process.

6.2.1 Synchronizer Synthesis. Related Work. Once the temporal constraints governing the relative sequencing of operations are explicitly available, synchronizers can be synthesized that enforce these constraints[13, 21, 19]. Such synchronizers are typically finite state machines that provide a model satisfying the given temporal and resource constraints. In [21], the design for the synchronizer proceeded from specifications in propositional temporal logic, and used the decision procedure described in [9] to generate a state machine. This was done in the context of an experiment to transform fragments of programs in a subset of Ada into hardware realizations cf.[2]; Anna [1] was augmented to enable synchronizer specifications using temporal logic[21].[9] Such an approach has two potential drawbacks: (1) The state machine generated by the algorithm in [9] may not exploit all of the available parallelism, and (2) the decision procedure involved is exponential.

In [13], the synchronization constraints are specified using path expressions. The generation of self timed

9. Anna, in turn, was an augmentation to Ada that allowed formal annotations to aid mechanical verification.

state machine implementations proceeding from data flow analysis done using compilation strategies is described in [2, 4]. The trade-offs involved in using these approaches is currently under investigation.

We now proceed to outline the generation of physical circuit layouts.

7. Other Experts

The translation from a structural description to a physical layout is achieved by using a group of experts in lower level tasks such as floorplanning and leaf cell layout CADRE[15]. The structural description may typically be subjected to local technology specific transformations[6]; these transformations can be based on an *algebraic model* of the technology being used e.g., structured logic arrays, NMOS etc. (cf. [16]). A set of algebraic rules of inference can be used to determine the "overall" geometric shape of the resulting networks (in the sense of the shapes of figures typically used to depict systolic arrays); such information can be used, for example, in guiding the floorplanning modules.

Leaf cell symbolic layout is provided by a rule-based layout tool called *Topologizer* [17]. This expert takes as input a description of transistor connectivity along with a set of constraints that define transistor sizing, external connection, area and aspect ratio. What it produces is a (CMOS) symbolic layout of the cell in a form suitable for processing by a virtual grid layout system such as *Mulga* [3].

7.1 Implementation Status

A skeletal prototype of the transformation system has been implemented in Prolog, and is being ported to a Lisp machine environment. The leaf cell layout expert has been implemented, while the floorplanning experts are currently being implemented. The technology specific transformation module is planned for but not yet implemented. The algebraic simplification/ theorem proving base runs on the VAX, and is implemented in Pascal[20].

8. Performance Criteria and Strategy Guidance

There are typically several design options available at each stage in the refinement of a design. One ramification of incorporating performance criteria into a specification is that the development strategy of a design can be guided by the desired performance criteria. We elaborate on some issues in this context, using the sort example for concreteness. Although our exposition below is couched in terms of area complexity, much of the discussion can be generalized to other performance dimensions e.g, response time, throughput.

The table below indicates some of the levels of hierarchy in the design of the linear network that implements the sort function, and the units of area that are used at each level of detail in a design.

Level	Element	Area Complexity (excluding routing)
1	N-element Sorter	O(N) * Area(Sorter-PE)
2	PE used in a N-bit Sorter	Area(N-bit Comparator) + Area(N-bit Multiplexor) + Area(N-bit Register)
3	N-bit Comparator (using bit slices)	O(bits) * Area(Comparator bit slice)
3	N-bit Multiplexor	O(bits) * Area(Multiplexor bit slice)
4	Comparator bit slice	O(gates)
4	Multiplexor bit slice	O(gates)
4	N-bit Comparator (not using bit slices)	O(gates)
5	Gates, Primitive Cells	O(transistors)
6	Transistors, wires	square mm.

Note that each of the above area expressions has a routing component which we have not explicitly indicated so as to avoid cluttering up the table. This routing area depends upon the number and nature of interconnections. Although an exact value for the routing area is typically not available until a complete physical layout has been done, it is possible to approximately account for this area by using an empirically determined scaling factor (> 1) by which the remainder of the area is scaled up. We conjecture that such an approximation might suffice in most cases for guiding the higher levels of the synthesis strategy.

When engaged in top-down refinement of the description of a sort chip, the area complexity is first expressed as a function of the number of elements in the list to be sorted (a number that is meaningful when talking about lists), and the unit of area is defined by the area occupied by one of the processing elements in the sort network. Note that this unit of area is not concretely estimatable at this point.

Each subsequent level of detail in the refinement of a design makes increasingly explicit the unit of area used by the preceding level of abstraction. At the lowest level (which we have here assumed to be gates or leaf cells from some library), the area is typically available in terms of square millimeters. Note that the dependence upon technology has been factored into this level, although it is possible to view technology dependence as a further level of detail in the development of a design.

When a particular design is developed to a level where the area is available in square millimeters, back substitutions can be performed to obtain a more concrete approximation to the area that will actually be taken up by a design. This estimate will still be "approximate", since there are some components of the actual circuit area e.g., routing area, that cannot be accurately estimated until the structural circuit description is converted to a physical implementation. However, as mentioned above, the area increase incurred due to routing may be approximated by an empirically determined function (a constant in the simplest case). Thus, a reasonably accurate estimate of the actual chip area can be obtained, and a decision can be made as to whether the estimate is consistent enough with the mandated performance specifications to warrant mapping the resulting structural description into a physical layout.

In general, it is possible to obtain an approximation to a performance metric, e.g., area, at any level of abstraction. However, if two different implementations of the same function use differing units of area then it is not possible to compare the two area estimates without descending to a level of detail where the area units are the same. We optimistically conjecture, however, that a comparative estimate can typically be obtained without exploring a particular design option to the fullest detail.

The table below indicates some of the design options during the initial stages of the sort chip design. In the table, DM denotes the function DeleteMax, PC(e) denotes the pin count of a list element, A and denote T the area and time complexity functions, BR denotes a bit-regiter, BC denotes a bit comparator, EC denotes an element comparator, N denotes the number of elements in the list, and b denotes the number of bits needed to represent each element of a list.

Module	Design Options	Area	Time	PinCount
Sortchip	Linear seq.	A(DM) + A(Control) + (N-1)*I/OBuffer	O(N)*T(DM)	4*PC(e)
	Linear pipe	O(N)*A(DM)	O(N)*T(DM)	2*PC(e)
DeleteMax	Rect. par	O(N)*A(EC)	O(N)*T(EC)	2N*PC(e)
	Linear (Serial out)	A(EC)	O(N)*T(EC)	2*PC(e)
	Linear (Parallel out)	A(EC)	O(N)*T(EC)	2*PC(e)
Comparator	Linear pipe	A(BC)+A(BR)	O(b)	2
	Rect pipe.	O(b)*A(BC)	O(b)*A(BC)	2*b+1
	Parallel	O(n^3)	O(1)	2*b+1

If we limit ourselves to just the possibilities listed in the table above, it is easy to see that the most "reasonable" implementation of a sort chip is one which uses a linear pipelined implementation where the elements of the list are streamed in sequentially; this implementation has area complexity O(N)*T(DM), time complexity O(N)*T(DM) and a pin count equal to 2*PC(e). If the pin count 2*b is acceptable, then the list elements can be streamed in in parallel. Otherwise, the elements themselves have to be input bit serially. If response time was critical, then (amongst the options above), the parallel implementation of an element comparator would have to be chosen; note that this choice results in an increase both in the pin count and the chip area. Other possibilities fall somewhere in between.

Of course, considerations different from those above might dictate the choice of one of the other options. For example, if it is required that the output of the list be available in parallel, then the "bubble sort" is not suitable, whereas if only a parallel input is available, then the linear array with parallel outputs might be a viable candidate.

An important point to note in the example above is that if the designs explored are guided by performance criteria, then the right choice can be made at each of these levels of abstraction, using only the approximate complexity metric available. It is obvious that the choices guided by the performance heuristics might not always be so fortuitous in the general case. We believe, however, that reasonably good choices can be made early on in the design, rather than having to wait until all of the details of a design option have been explored. Furthermore, while it is obvious that leeway in design available at each stage of development must be exploited as best as possible, it is important to underline the fact that a much bigger leverage is typically available in the earlier stages of design. As in the case of software design, a bad initial choice cannot generally be remedied by later optimizations.

9. Summary

We have presented an algebraic framework that underlies the design of an expert system for VLSI design, and illustrated the use of provably correct transformation schemas that map behavioral specifications into computational networks. We believe that the ability to experiment with alternative system designs at higher (architectural) levels that is provided by such a framework enables the concurrency inherent in a problem to be exploited in ways that may not always be possible in the more traditional forms of optimizations that are typically attempted on a *fixed* system architecture.

One of our main motivations for using an algebraic basis for this development was to attempt to provide a cohesive framework for reasoning about various aspects of the overall VLSI design process; related efforts are discussed in[18, 7]. It is therefore appropriate to summarize here some of the key facets of design where such an algebraic basis is applicable, although this paper does not elaborate on all of these issues.

- The specifications of both functions and networks that implement these functions is algebraic.
- The transformations that map between functional specifications and networks that implement are algebraic, and the proofs of correctness of these transformations are done using algebraic manipulations.
- The manipulations of functional specifications that are sometimes needed to enable implementation schemas to be applied are also algebraic, and use the (algebraic) definitions of the functions.
- The computation of the performance metrics are defined algebraically.
- System timing and synchronization information is specified using axiomatically characterizable logical primitives.
- Lower levels of abstraction in a circuit representation may be modelled using these algebraic primitives e.g., structured logic arrays [16] and MOS networks; algebraic equation solving techniques can be used for supporting simulation and verification tools at these levels.

Needless to say, we have only scratched the surface of what must be done in these areas. Nevertheless, it has been our experience that attempting to base expert systems on a sound footing is a worthwhile task. This belief was underlined by the fact that the initial statements of some of the transformations described by designers were found to be inaccurate (or incomplete) in some cases, and this was not discovered until the proofs of those schemas were attempted! The designs produced in isolation by some of the automated experts compare favorably to that of human design (within 10-20 percent). While it is premature to generalize from a few examples, the outlook is certainly promising.

Appendix I: Definition of the type List

```
Type List(Element)(Integer)

Syntax
NewList: () → List
Add: List, Element → List
Delete: List, Element → List
Sort: List → List
Size: List → Integer
Max: List → Element
 ...

Semantics

The parameter type Element must have comparison operators <, and > defined
on it. In addition, it is presumed that it has two elements -∞ and +∞
that are the least and greatest possible elements.

Max(NewList) = - ∞
Max(Add(l,x)) = if x > Max(l) then x else Max(l)

Delete(NewList,x) = NewList
Delete(Add(l,x), x') = if x=x' then l else Add(Delete(l,x'),x)

Sort(NewList) = NewList
Sort(Add(l,x)) = Add(Sort(Delete(Add(l,x),Max(Add(l,x)))), Max(Add(l,x)))

Size(NewList) = 0
Size(Add(l,x)) = Size(l)+1
End List;
```

Appendix II: Composition Primitives

The equational semantics of some of the system description primitives is given below. For convenience, we have used infix notation. Let X, Y denote concurrent systems, and A, B primitive actions (e.g., I/O); nil denotes termination.

$$
\begin{array}{ll}
x + y = y + x & \text{(CC)} \\
x + (y + z) = (x + y) + z & \text{(CA)} \\
x + x = x & \text{(CI)} \\
(x + y) \sim> z = x \sim> z + y \sim> z & \text{(CSD-1)} \\
x \sim> (y \sim> z) = (x \sim> y) \sim> z & \text{(SA)} \\
x + \text{nil} = x & \text{(CID)} \\
\text{nil} \sim> x = \text{nil} & \text{(SZ)} \\
x|y = y|x & \text{(CM-C)} \\
x|(y|z) = (x|y)|z & \text{(CM-A)} \\
x|\text{nil} = \text{nil} & \text{(CM-Z)} \\
x||y = x<|y + y<|x + x|y & \text{(PAR)} \\
A<|X = A \sim> X & \text{(SPP)} \\
(A \sim> X) <| Y = A \sim> (X||Y) & \text{(SPS)} \\
(X + Y) <| Z = X \sim> Z + Y \sim> Z & \text{(SPC)} \\
(a \sim> x) \mid b = (a|b) \sim> x & \text{(CMP1)} \\
a \mid (b \sim> x) = (a|b) \sim> x & \text{(CMP2)} \\
(a \sim> x) \mid (b \sim> y) = (a|b) \sim> (x||y) & \text{(CMP3)} \\
(x+y)|z = x|z + y|z & \text{(CSD1)} \\
x|(y+z) = x|y + x|z & \text{(CSD2)} \\
A - P = \text{if } A \in P \text{ then nil else } A & \text{(ABS1)} \\
(x+y) - P = (x-p) + (y-p) & \text{(ABS2)} \\
(x \sim> y) - p = (x-p) \sim> (y-p) & \text{(ABS3)}
\end{array}
$$

Intuitively, the composition (|) axioms depend upon whether the actions are input/output.

$$
\begin{array}{l}
(p \leftarrow X) \mid (p \rightarrow X1) = (X{=}X1) \\
(p \leftarrow X) \mid (q \rightarrow Y1) = \text{nil} \\
(p \leftarrow X) \mid (q \leftarrow Y) = \langle p \leftarrow X, p \leftarrow Y \rangle \\
(p \rightarrow X1) \mid (q \rightarrow Y1) = \langle p \rightarrow X1, p \rightarrow Y1 \rangle
\end{array}
$$

This extends to parallel inputs and outputs. Thus, for example,

$$\langle p \leftarrow X, q \leftarrow Y \rangle | \langle p \rightarrow X1, q \rightarrow Y1 \rangle = (X{=}X1, Y{=}Y1)$$

In general,

1. Opposite polarity transactions of unrelated ports yield nil.
2. Like polarity actions at unrelated ports are concurrent.
3. Opposite polarity actions at complementary ports result in a substitution.

Appendix III: Transformation Schemas

Schema DP. Consider the function $F : T, X \rightarrow T$ defined by:

$$
\begin{array}{l}
F(\text{NewT},y) = \text{NewT} \\
F(c(t,x),y) = \text{if } P(x,y) \text{ then } t \text{ else } c(F(t,y),x)
\end{array}
$$

where P has the functionality $P{:}X, X \rightarrow \text{Boolean}$. Consider now the behavior

$$FM\{L{:}T, R{:}T\} = L \leftarrow \langle t,y \rangle \sim> R \rightarrow F(t,y) \sim> FM\{L{:}T, R{:}T\}$$

This behavior can be implemented using a module FModule{L:X, R:X} defined by:

$$
\begin{array}{l}
\text{FModule} = L \leftarrow y \sim> \text{FModule1}(y) \\
\text{FModule1}(y) = L \leftarrow x \sim> \ \text{if } x{=}\text{eos then } R \rightarrow \text{eos} \sim> \text{FModule} \\
\qquad\qquad\qquad\qquad \text{elseif } p(x,y) \text{ then FModule1(eos) else } R \rightarrow x \sim> \text{FModule1}(y)
\end{array}
$$

with the following I/O Representation Functions:

I: T→ Stream(X ∪ {eos})
O: Stream(X ∪ {eos})→ T

I(NewT) = eos
I(c(t,x)) = Concatenate(Seq)(Append(EmptySequence,x),I(t))

O(EmptySequence) = NewT
O(Append(s,x)) = Concatenate(T)(c(NewT,x),O(s))

This says that the input to FModule is a streamized version of the input t to FM in "reverse order", i.e., c(...c(NewT,x1),...),xn) is input as the stream <xn,...,x2,x1,eos> where "eos" is the end-of-stream designator. The functions Concatenate(Seq) and Concatenate(T) are analogous to the list append functions, and we do not define them formally here. The parameter in the definition of FModule1 represents an internal state that is initialized with the first input on the port L.

The functional behavior of the schema can be characterized by:

L→ I(t) | FModule = O(R→)=F(t,y) ~ > FModule

which says that if the encoded input is presented at the port L, then the output representing F(t,y) will later be available at the port R. ■

Proof of Correctness of Schema DP. Omitted.

Schema DT. Consider the function F : T, X→ T defined by

F(NewT,y) = NewT
F(c(t,x),y) = if P(x,y) then t else c(F(t,y),x)

where P has the syntax P:X,X→ Boolean. Then the behavior

FM{In1:T,In2:X,Out:T} = <In1← t,In2← y> ~ > Out→ F(t,y) ~ > FM{In1:T,In2:X,Out:T}

may be implemented using the I/O Representation Functions

I: T→ Tuple(X)(N)
O: Tuple(X)(N)→ T

I(NewT) = EmptyTuple
I(c(t,x)) = Append(I(t),x)

O(EmptySequence) = NewT
O(Append(v,x)) = if x='*' then O(v) else C(O(v),x)

by FMODULE defined by:

FMODULE {Top[1..N], Right, Left, Bottom[1..N]} =
PM{Top[N], Right, PMLeft, Bottom[N]} |
FMODULE {Top[1..N-1], PMLeft, Left, Bottom[1..N-1]}

or, equivalently:

FMODULE(N) = pipeline(1,N)(PM,connect([Right--Left]),hide({Right,Left}))

The submodule PM implements the function P and has the behavior:

PM = <Top← t, Right← r> ~ >
if p(t,r) then <Bottom→ *, Left→ *> else <Bottom→ t, Left→ r>~ > PM

It can now be shown that the module FMODULE exhibits the following behavior:

<FMODULE.Top→ I(t), FMODULE.Right→ y> | FMODULE =
<O(FMODULE.Bottom→)=F(t,y), O(FMODULE.Left→)=H(t,y)> ~ > FMODULE ■

References

[1]. Krieg-Bruckner, B., Luckham, D.C., von Henke, F.W., Owe, O. , *(Draft) Reference Manual for Anna, A language for Annotating Ada Programs* , Stanford University, Dept of Computer Science, 1982 ().

[2]. Organick, E.I., Keller, R. M., Lindstrom, G., Smith, K.F. and Subrahmanyam, P.A., *Transformation of Ada Programs into Silicon. Third SemiAnnual Technical Report,* University of Utah (April 1983).

[3]. Weste, N. and B. Ackland, "A Pragmatic Approach to Topological Symbolic IC Design," pp. 117-130 in *Very Large Scale Integration,* ed. John P. Gray, Academic Press (1981).

[4]. Tony M. Carter, "ASSASSIN: An Assembly, Specification and Analysis System for Speed-Independent Control-Unit Design in Integrated Circuits Using PPL," UTEC-82-103, University of Utah, Department of Computer Science (June 1982).

[5]. B. Cohen, "On the Impact of Formal Methods in the VLSI Community," pp. 469-481 in *Very Large Scale Integration (VLSI 83),* ed. F.Anceau and E.J.Aas, North-Holland, Amsterdam (August 1983).

[6]. Darringer,J.A., D.Brand, J.V.Gerbi, W.H.Joyner, and L.Trevillyan, "LSS: A System for production logic synthesis," *IBM Jour.of Research and Development* **28**(5) pp. 537-545 (September 1984).

[7]. M. Gordon, "A Very Simple Model of Sequential Behavior of nMOS," in *Very Large Scale Integration,* ed. John P. Gray, Academic Press (1981).

[8]. Hayes, A.B., "Stored State Asynchronous Sequential Circuits," *IEEE Trans. on Computers* **C-30** pp. 596-600 (August 1981).

[9]. Z. Manna and P. Wolper, "Synthesis of Computing Processes from Temporal Logic Specifications," *Proceedings of the Workshop on Logics of Programs, Yorktown Heights, NY,* Springer Verlag LNCS, (1981).

[10]. Z. Manna and A. Pnueli, "Verification of Concurrent Programs: The temporal framework," pp. 215-273 in *The Correctness Problem in Computer Science,* ed. R. S. Boyer and J. S. Moore, Academic Press, N.Y. (1981).

[11]. R. Milner, *A Calculus for Communicating Systems,* Springer Verlag Lecture Notes in Computer Science (1980).

[12]. R. Milner, *Calculii for Synchrony and Asynchrony,* Technical Report, Department of Computer Science, University of Edinburgh (April 1982).

[13]. T.S.Ananthatraman, E.M. Clarke, M.J.Foster, B. Mishra, "Compiling Path Expressions into Circuits," *Proc. 12th Ann. SIGACT/SIGPLAN Symposium on Principles of Programming Languages,* pp. 191-204 (January 1985).

[14]. B. Moszkowski, *Reasoning about Digital Circuits,* PhD Thesis, Stanford University, Dept. of Computer Science (1983).

[15]. B.D.Ackland, A.Dickinson, J.R.Ensor, J.D.Gabbe, T.London, C.Porier, P.A.Subrahmanyam, H.Watanabe, "CADRE - A system of Coperating VLSI Design Experts," *1985 IEEE International Conf. on Computer Design: VLSI in Computers,* pp. 99-104 (October 1985).

[16]. Rajopadhye, S. and P.A.Subrahmanyam, "Formal Semantics for a Symbolic IC Design Technique: Examples and Applications," *Integration: The VLSI Journal,* (3) pp. 13-32 (March 1985).

[17]. Kollaritsch, P.W., N.H.E.Weste, "TOPOLOGIZER: An Expert System Translator of Transistor Connectivity to Symbolic Cell Layout," *VLSI Design,* pp. 62-66 (August 1984).

[18]. Cardelli, L. and G. Plotkin, "An Algebraic Approach to VLSI Design," pp. 173-182 in *Very Large Scale Integration,* ed. John P. Gray, Academic Press (1981).

[19]. K. Ramamritham and R.M. Keller, "Specification and Synthesis of Synchronizers," *Proc. Intl. Symp. on Parallel Processing,* pp. 311-321 (August 1980).

[20]. Subrahmanyam, P.A., "The Architecture of SYNAPSE: An Expert System for VLSI Design," Bell Labs Technical Memorandum (January 1985). To appear in IEEE Computer

[21]. Subrahmanyam, P.A., *From Anna+ to Circuits: Automating the Synthesis of Hardware Synchronizers from Temporal Logic Specifications,* University of Utah, Department of Computer Science (March 1982). Internal Report

[22]. Subrahmanyam, P.A., *An Algebraic Basis for VLSI Design,* Dept. of Computer Science, University of Utah (October 1982). Draft of a Research Monograph

[23]. Subrahmanyam, P.A., "Transformation Techniques for Systolic Arrays," Bell Labs Technical Memorandum 84-11354-09 (December 1984).

Formal Aspects of VLSI Design
G.J. Milne and P.A. Subrahmanyam (editors)

SPECIFICATION AND VLSI DESIGN

Graham Birtwistle, Jeff Joyce, Breen Liblong (+),
Tom Melham (++), and Rick Schediwy,

Computer Science Department, University of Calgary,
Calgary, Alberta, Canada T2N 1N4.
(+) now at Silicart, Montreal, Canada.
(++) now at Cambridge University, U.K.

ABSTRACT

We describe research into specification-based VLSI design underway at the University of Calgary. Our long term research goals are directed towards building a specification-based design environment (EDICT) to support an iterative, hierarchic design methodology. Our current research has three aspects: the SHIFT high level design capture format (completed); gaining experience in verifying large designs (underway); and building a specification library. In this paper we describe work in progress on two large proofs. The first is for the elimination unit of a local area network device, for which the proof is well underway. The second project concerns the specification driven design of Landin's SECD machine and is just beginning. To set the context for this work on verification, we start by giving partial descriptions of EDICT and SHIFT to show how they use specifications.

INTRODUCTION

In EDICT [2] we picture a design under construction as a tree, every node of which has a specified functional behaviour. If we compose together the behaviours of a node's immediate offspring, we can check at once to see whether or not their composition agrees with their parent's specification. If so we have a correct refinement of the design; if not we know at the earliest possible stage that we are wrong. By deductive argument, we can show the correctness or otherwise of a complete design (within the limits of the proof model - we work at the register transfer level and do not include electrical effects). Specifications are expected to play an increasingly important role in VLSI design, especially for military, industrial control and medical applications where the guarantee of design soundness is crucial. They also assist the process of splitting designs amongst teams in such a way that the function of the whole chip need not be revealed in its entirety to any sub-contracted party.

The output from EDICT is in a high level VLSI design exchange format called SHIFT [15, 16] which captures and retains the structure of the design as well as its low level detail. The SHIFT description of a design follows the behavioural hierarchy exactly. A SHIFT description consists of several views (e.g. behaviour, constraints, electrical properties, geometry, structure, ...). These views have been chosen to enable SHIFT to interface easily to current (and future) lower level tools such as SPICE, MOSSIM, timing verifiers and plotters. These special purpose tools can be used to fill out those views in a SHIFT definition that are not derivable from the behaviour. For example, the electrical-properties view can be filled out via SPICE for leaf cells, and higher level simulators can be used to check timing constraints (e.g. port to port signal timings). Other standard packages will provide plots and foundry tapes etc. SHIFT supplies composition tools for all views so that large designs can correctly constructed automatically from validated library elements. SHIFT is also used as the design data base language for validated designs. Like EDIF [3, 4] to which it bears a very close resemblance, SHIFT may also be used to port designs between workstations, to different fabrication lines, and to put designs in the public domain.

The latter sections of this paper outline progress on two substantial register transfer level proofs of correctness. The first proof will demonstrate the correctness of a design for the elimination unit of a local area network box. The second is concerned with an SECD machine [9, 13]. Both proofs are being effected mechanically using proof assistants. We worked with LCF_LSM [5] for the elimination unit proof which was started in 1983, but swapped to HOL [6] for the SECD specification work.

OVERVIEW OF EDICT

The key elements in EDICT [2] are support for a design methodology, the use of proof techniques to ensure that designs meet their specification, and SHIFT.

One method of controlling the complexity of VLSI designs is to reduce the design space by imposing a design methodology. We have chosen to work with hierarchies where a design is constructed as a sequence of levels each lower one a proven consistent refinement of the one above. Because today's design becomes tomorrow's building block, we also arrange to collect validated designs in a library from whence they can be offered, perhaps semi-automatically, to future designers. In this way, we are accommodating and expanding the standard cell approach. Within the EDICT system we create large cells automatically by composition, and will be looking into the possibilities for system-polished designs (as with TOPOLOGIZER [12]).

Designs in progress are represented in EDICT by a tree of nodes. A simple example presented in [8] considers the implementation of a sampler which inputs a serial bit stream and outputs a 4-bit word. The output value at time t is #0000 if the sample line is low; if the sample line is high, it is the concatenation of the input bits at times t-3 through t. We can decompose the sampler first into a bus and a shifter, then the shifter into four slices, each slice into a delay an inverter and a nor gate, and finally implement the delay unit by a D-type flip-flop. This design decomposition is shown in figure 1.

If we design from the top down, the single node at level 1 represents the complete design, its offspring at level 2 the floor plan elements, and so on. Initially, we will have the specification of the design's behaviour (which expresses output signals in terms of current and previous values on its input lines), together with some performance expectations and area limitations expressed as constraints. Our first task is to partition the design into a set of components whose behaviour when connected together conforms to the specification of the design at level 1, and will satisfy its constraints. We specify the behaviour of each floor plan element in turn and write down its local constraints, if any. Then we compose the behaviours of the floor plan elements together and deduce the behaviour of this first design refinement. If the aggregate behaviour can be shown to be equivalent to that of the design at level 1, then we have a valid implementation (subject to the constraints being satisfied) and can recursively apply the same technique to each floor plan element in turn. If the specifications do not match, then we know at the earliest possible moment that we are on the wrong track.

As a design develops, the nodes become simpler and simpler until they can be implemented at once by a leaf cell editor or already exist as library elements. End nodes in the design tree are called leaf cells (e.g. bus, not, nor, D-type

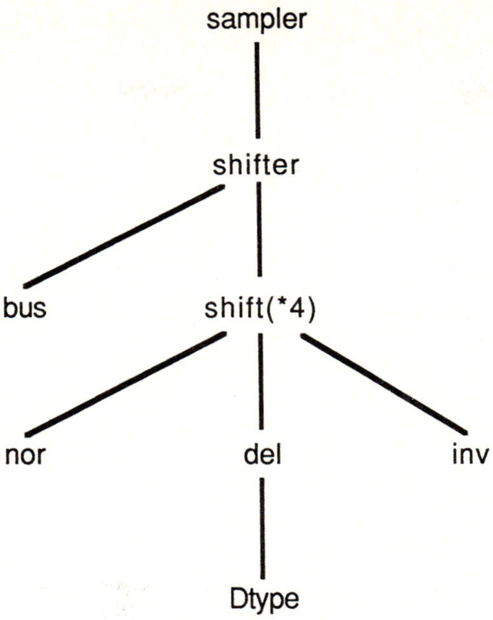

Figure 1 Sampler hierarchy

above) and intermediate nodes are called composition cells (e.g. shifter, slice, del above). Leaf cells are fully developed and, besides their inherited behaviours and ports, will have synthesized attributes such as structure, layout, schematics, and known power consumption, delay, and area.

New leaf cells will be hooked up to SPICE, MOSSIM, timing verifiers, design rule checkers and layout tools etc for complete validation and to ensure that they satisfy their constraints. When templates for all classes of leaf cell instances are known in full, the dialog with the user is suspended, and an automatic design composer takes over. The composer has built in composition rules for all template attributes. It composes the structure and layout of offspring and fills in the attributes of their parent. It will do the same for power requirements, delay, and area. We now work our way back up the design tree applying composition rules to fill out the composition cell templates and to check that they satisfy their constraints. A cell whether it is a leaf cell or a composition cell is not complete until all its inherited and synthesized attributes have been filled out and its local and imposed constraints satisfied. Once a template is complete and verified, it is included in the SHIFT library and will be available to future designers.

The automatic composer back end effects the compositions remaining (other than the behaviours). If a design fails to meet its constraints, then EDICT can help at the next iteration. Templates from the previous iteration should contain better estimates for critical values than were available on the previous unsatisfactory design attempt. These can be taken into account and may lead to global optimisations, and floor plan alternatives or architectural choices being investigated early on in the next iteration.

Thus the overall structure of EDICT is an interactive design capture front end and and automatic composer. The front end constructs the design tree by requesting specifications from the designer, composes specifications and assists in the arduous task of checking whether or not two specifications are the same. The front end fills in only the specification portion of the design template. The composer extracts completed templates from the library and composes them to complete the design.

THE SHIFT DESIGN CAPTURE LANGUAGE

SHIFT is described in [15, 16]. A prototype for flexible NMOS and CMOS cells was implemented in 1984 and has been enhanced since. SHIFT will be used in EDICT to capture the various descriptions of a circuit in a consistent manner, and support the development of the design through incremental refinement. SHIFT retains the structure and hierarchy put into a design, supports composition, and serves as a library format.

SHIFT bears a strong resemblance to EDIF [3, 4] (Electronic Design Interface Format) which is being developed as a language for design exchange between designers and as a foundry interface language, but not as a design data base language. SHIFT and EDIF were developed quite independently.

SHIFT is embedded in LISP and the syntax retains the same flavour. A design is specified by leaf cells and composition cells. All cells are rectangular with a boundary on which ports are placed. Cells are composed by abutment in horizontal and vertical directions. The composed cells are stretched to connect adjacent ports.

A leaf cell is specified by a name, a list of ports, geometry, structure, electrical properties, a minimum bounding box, behaviour, and a set of constraints.

```
(defcell cell_name
        (ports....................)
        (geometry.................)
        (structure................)
        (electrical properties....)
        (mbb......................)
        (behaviour................)
        (constraints..............)
```

Ports are specified along the north, south, east and west walls of the cell, and it is only through abutment of these external ports that instances of cells are connected in any of the views. This view is derived from the behaviour and constraints fields, and is used in floorplanning, placement, and routing stages.

The geometrical view is specified as a series of geometric patterns on distinct layers corresponding to the fabrication masks. A range of operators is provided for specifying and transforming polygons, boxes, wires, and contacts. Leaf cells are made stretchable by specifying constraints between the ports. Any geometric coordinates specified relative to the ports are determined when the cell is instantiated. The actual connection of layers through the abutting ports of the components of a composition cell are checked separately by a geometric design rule checker.

The structural view is specified as a collection of components and a netlist. Components are named instances of transistors, resistors and capacitors. Connections are then specified between the gate, source and drains of the transistors, the terminals of the resistors and capacitors, and the ports. We will be investigating how to derive this view from the behavioural view automatically.

The electrical properties field supplies additional information about the components, such as relative strengths of the transistors, relative sizes of the capacitances specified as functions dependent on the port values in order to facilitate performance simulation, pin to pin timings and power estimates. SPICE is used for leaf cells. We will be using algorithms similar to those of Lin [17] and Trimberger [25] respectively to compose electrical properties and optimise delays and power. We are also incorporating restoring logic checks following Mead and Rem's algorithms [24].

The minimum bounding box, mbb, is derived from the geometry view and is used in floorplanning, placement and routing.

Constraints may be specified in the design of a composition cell for any view. For example, the physical dimensions of the resulting identity cell could be specified through constraints on the composition cell's ports that must be met in the instantiation of the component cells. Similarly we could specify that the list of connections from the composition of the three component cells must match some structural specification of the count cell. Finally a behavioural constraint could be that some specified behaviour of the composition cell must match the behaviour derived from the the component cells.

Composition cell definitions require two fields - the composition itself plus any additional constraints that may be made. For example, a 4 by 4 array of SHIFT register cells can be constructed from a single SHIFT cell by

```
(defcomp SHIFT4ROW (> (rep 4 SHIFT)) nil)
(defcomp SHIFT4by4 (^ (rep 4 SHIFT4ROW)) nil)
```

where > and ^ are the abut east and abut north operators and 4 the repetition factor. No extra constraints are specified. If empty, other views are filled out automatically by composition If not empty they may be considered as specifications to be met. The composition rule specifies the patterns for the masks from the patterns of its constituents. It also specifies the structure as two subgraphs being joined together at adjacent ports. The behavioural description is given by the composition of the behaviours of the components, where abutting ports are identified. And so on. Nothing extra in the specification in any of the views is introduced by the composition rule that is not derivable from the component specifications.

The behavioural view will be specified in HOL [6]. One current research topic is to find out just how much can be deduced from the specification alone. For example, it is quite straightforward to generate hierarchical simulators automatically from typed specifications (these will be used to give confidence in the correctness of specifications as design work progresses) and it is also straightforward to produce first cuts at floorplans automatically from the wiring (just using the typed ports) in the manner of Nixon [22].

TOWARDS THE VERIFICATION OF THE ELIMINATION UNIT

The flooding sink is a local area network proposed, designed, and analysed at the University of Calgary by a team led by Brian Unger [23]. The design permits specially tailored nodes to be linked together in an arbitrary fashion and is suitable for use as a real time local area network requiring a very high bandwidth.

Attached to each node is a host through which users enter and receive messages. Each node contains an arbitrary number of connections to other nodes numbered [0..n]. A connection between nodes consists of a full duplex link. One link (which we number 0 for textual convenience) is reserved for communication with the host.

Messages circulating in the network contain a source address, a serial number, and a destination address each of length 8 bits. The 16 bit field 'source address + serial number' is taken as the message identifier. When a host sends a message, a copy is routed into its node and from there broadcast along each of the output lines [1..n], but not back to the host. Messages received are buffered separately for each input port. In a broadcast system, the same message may reach a particular node several times via different routes. Incoming messages are thus first checked to see if they have been received before. If so, they are discarded. If not, they are added to the list of messages received and will be rejected if received again later. The analysis given in [23] shows that overall system performance depends heavily the ability to accept new messages and reject repeats quickly. The elimination unit is a hardware response to that problem and it is with its implementation that we are concerned.

The elimination unit polls each input buffer in turn, looking for a request from the buffer to check an incoming 16-bit message ID. When a request is found, the elimination unit checks the message ID against a list of the last 256 identifiers to reach this node by using the message ID as an index into a 64K by one-bit memory, lookup. This memory returns true if the identifier is one of the last 256 received or false if it is not. If true is returned, the message is not to be broadcast via the output ports and the eliminator responds to the request with keep=false. If false is returned by the memory, the eliminator responds keep=true so that the message is kept for broadcasting to the other network nodes. In this case, the eliminator updates lookup by setting the bit indexed by the incoming ID to true. In order to ensure that lookup contains only 256 bits which are set, the elimination unit has a 256 long FIFO in which the last 256 ID's to arrive are stored in order. When lookup is updated, the bit indexed by the oldest ID in the FIFO is set to false, so that lookup will contain only 256 set bits. The FIFO is implemented using a 256 by 16 bit memory, lastids, and a pointer, oldest, which is used to point to the oldest ID in the queue. By using two memories as outlined above, the elimination unit does not have to sequentially search a list of previously seen IDs and therefore is able to respond very quickly to look up requests. Pseudo-code for the elimination algorithm is given below:

```
ELIMINATION UNIT:

  for k := 0 to n do
    if buffer[k] requests an ID look up then
      if lookup[new_ID] then signal keep=false else
        {! set look up memory right;
         lookup[new_ID] := true;
         lookup[lastids[oldest]] := false;
         ! add new ID to FIFO;
         lastids[oldest] := new_id;
         oldest := oldest+1 mod 256;
         keep := true;
        }
     wait until buffer[k] drops request;
    repeat;
```

A prototype of the elimination unit was designed and breadboarded by Rick Schediwy and the proof undertaken by Tom Melham at the University of Calgary. Implemented in bipolar, the prototype consists of 37 SSI, MSI and LSI chips ranging in complexity from 4-input NAND gates to AM2910's. The proof has proceeded from the bottom-up. At the time of writing, full proofs have been completed for all the floor plan elements and what remains is the matching of their composition to that of the high level specification. Details of the proof of this device are beyond the scope of this paper. Even in its present incomplete state it covers 9000 lines of LCF-LSM. Instead of giving details of the proof at all levels, we will follow down one branch of the full proof tree from the top level to the bottom.

At the top level we view the elimination unit as a device with three ports and six states (see figure 2).

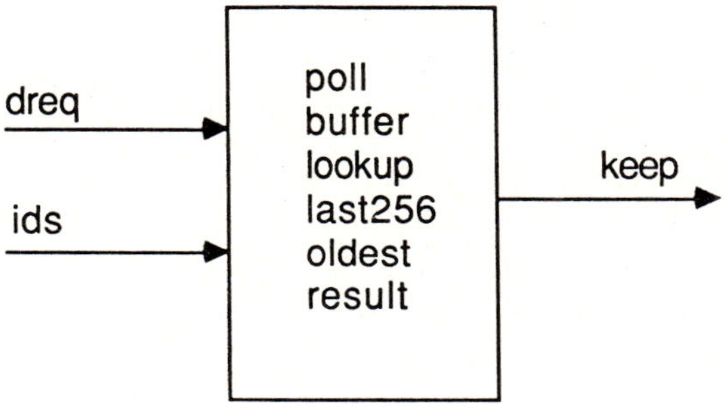

Figure 2 Top level view of the elimination unit

The ports are named dreq, ids, and keep. Requests to look up identifiers are received on the 8-bit bus, dreq, which is polled by the elimination unit. The input port, ids, is a list of 8 16-bit busses from which the incoming message IDs are read. The output line, keep, is a one-bit flag which indicates that an identifier has been seen previously (keep=false) or has not (keep=true).

The elimination unit has six state variables: poll, buffer, lookup, lastids, oldest, and result. The variables, lookup and lastids, are the 64K by one-bit look up memory and the 256 by 16-bit FIFO, respectively and the state variable, oldest, is the 8-bit FIFO pointer. The three-bit register buffer stores the number of the input buffer currently being polled. The result of the ID look up is stored in the one-bit state, result. The flag, poll, indicates which of two states that the device is in. When poll=true, the elimination unit is at the beginning of a polling cycle, looking for a request from the input buffer whose number is given by buffer. In this state, the keep output has the value true. When poll=false, the result of the ID look up is on the output and the elimination unit is waiting for the requesting input buffer to drop its look up request.

The specification of the elimination unit in LCF_LSM is:

```
ELIM(poll, buffer, lookup, last256, oldest, result)
  ==
dev {dreq, ids, keep}.
{keep = (request and not poll -> not result | T)};
ELIM( (request -> F | T),
      (request -> buffer | INC3 buffer)
      (request and poll and (not res)
            -> STORE16 id (#1) (STORE16 oldid (#0) lookup) | lookup),
      (request and poll and (not res) -> STORE8 oldest id last256 | last256),
      (request and poll and (not res) -> INC8 oldest | oldest)),
      (request -> (poll -> res | result) | F))
    )
```

where request is the input request from the current input buffer (formally an abbreviation for REQUEST buffer dreq which returns the value of dreq[buffer], the 8-bit request bus indexed by the current input buffer), ids returns the identifier belonging to the input buffer currently being polled from the list of input identifiers (formally EL (VAL3 buffer) ids), and res returns the lookup memory indexed by the incoming identifier (formally this value is given by BOOL1(FETCH16 lookup id).

This specification states that the device ELIM, with the six state variables mentioned above, has the behaviour given by the 'dev' expression on the right hand side of the equation. The first part of the dev expression is just a list of the device's external ports: dreq, ids and keep. The second part of the dev expression gives an equation for the output line, keep. The output equation states that if the device is not in the polling state (i.e. not poll) and the incoming request for ID look up is still active then the output will be given by the negation of the stored result of the look up, result. Otherwise, the keep output will have the value true.

The third part of the dev expression gives the 'next state' expression for each of the state variables. The first of these gives the next state for poll. If there is a look up request then the next state of poll will be false, otherwise poll will become true. Thus, whenever there is no incoming look up request, the elimination unit goes into polling mode and, whenever the incoming request is active, the elimination unit goes into output mode. The next state expression for buffer specifies that, when there is no request, buffer is incremented MOD 8 (by INC3) so that the elimination unit will go on to poll the next input buffer. Otherwise, buffer remains the same. The next states of lookup, ids and oldest all depend on the expression request and poll and (not res). When this expression is true, the device is in polling mode, there is an identifier look up request and the incoming identifier has not been seen before. In this case, the three states lookup, ids and oldest are all updated to reflect the fact that the new identifier has now been seen. When there is no request, the next state of result is the value false. Otherwise, the next state of result depends on the value of poll. When poll is true, the next state of result is the result of the identifier look up, res. When the elimination unit is not in polling mode (poll=false) the value stored in result stays constant.

The elimination unit is implemented using eight components START, IDBUS, RAMA, RAMB, COUNTERS, RESULT, DRIVERS and UNCNTL wired together as shown in figure 3. The LCF_LSM specification of the implementation is:

```
ELIM_IMP(start,device,lookup,last256,count,lastinc,buffer,
         result,uinstr,reg,adder,stack,sp,ucode)
  ==

   [| START start
    | IDBUS rn[en=det]
    | RAMA lookup ]
    | RAMB last256
    | COUNTER(count, lastinc)
    | RESULT(buffer, result)
    | DRIVERS
    | UCNTL(ucode, uinstr, adder, sp, stack, device)
   |]
   hide {adata,addr,ainit,awe,bsel,bwe,clear,drivers,
         init,nffl,nffh,rado,ramb,reply,req,run}
```

which lists the composing elements, how they are wired together including any wire renamings, and also lists the internal (hidden) wires.

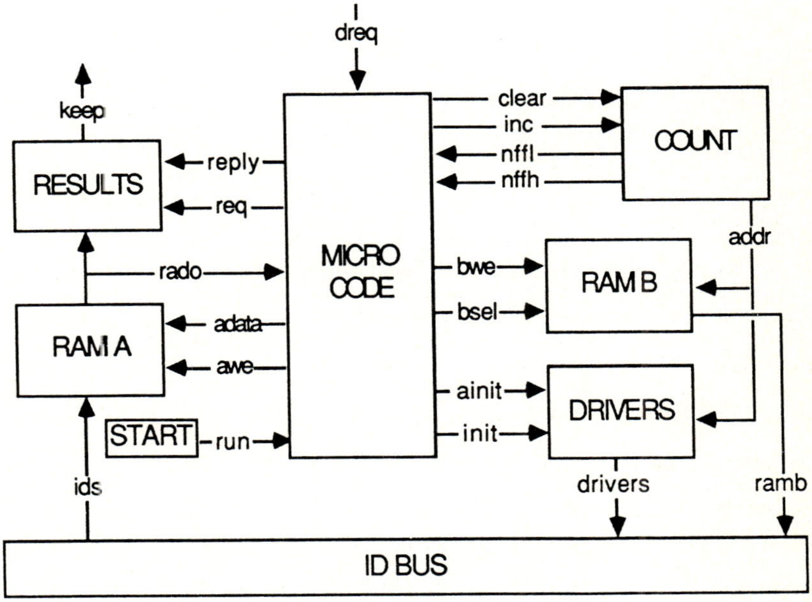

Figure 3 Elimination unit floor plan

START is used for proof purposes to simulate initialization from power-up. When its state goes from false to true, the initialization sequence will begin. When its state remains constant at true, the device is in regular operating mode. At this time, we have not worked on the correctness of the initialization cycle. IDBUS is the 16-bit bus for incoming identifiers which serves two purposes. It is used 1) during the initialization cycle to address RAMA from the counters, and 2) to address RAMA from the contents of RAMB when clearing the oldest ID bit in lookup. RAMA is the 64K by one-bit look up memory (a 64K by one-bit RAM, with state lookup. RAMB is the 256-item FIFO memory, with state last256. COUNTER is a 16 bit counter which serves two function. During the initialization sequence, it generates addresses for clearing RAMA and RAMB. Otherwise, the LOW 8 bits are used as the pointer into RAMB, (the FIFO pointer oldest). COUNTER has two outputs, nffl and nffh which indicate when the low and high bytes (respectively) have the value #11111111. This data is used by the controller during initialization. RESULT stores the look up result and generates the keep output. One of its states, result, is the result state of the specification. DRIVERS is used to control what is put onto the IDBUS from internal sources, and is discussed further below. UCNTL is the controller for the whole device; it uses microcode and an AM2910 microcontroller to generate control signals.

The proof is quite large (several thousand lines) and even its specification is too large to give in this paper. The specifications and proofs of all these top level components is now complete. What remains is the final step of showing that they compose to the original specification.

In what follows we trace the implementation of DRIVERS. DRIVERS controls what is put onto the id bus from the counters. The line drivers goes to the bus, the line addr comes from COUNTER. The specification of DRIVERS is:

```
DRIVERS
  ==
dev {ainit, init, addr, drivers}.
{drivers = (init -> FLOAT16 | (ainit -> MK_TRI16 addr
                 | MK_TRI16(MERGE_BYTES(#00000000, HIGH_BYTE(addr)))))};
DRIVERS
```

DRIVERS has three inputs ainit, init, addr and one output line drivers. When init is true, DRIVERS is disabled and the value put onto the bus is a 16-bit floating value FLOAT16. When init is false, DRIVERS is enabled and is being used to put addresses onto the bus in order to address RAMA and RAMB for initialization. When ainit is true, we are addressing RAMA and the address addr is put directly onto the bus. MK_TRI16 is an explicit type conversion which changes the value of addr from a 16-bit type which only includes 16-bit integers to a type which also has the floating value. When ainit is false, we are addressing RAMB using the high byte of the counter address, addr. The low byte output consists of a padding of 0's.

DRIVERS was implemented as the composition of seven primitive units.

```
DRIVERS_IMP
  ==
[| ZERO4        rn[out=zero4]
 | SPLIT_BYTE   rn[low_nibble=low4; high_nibble=high4; bytein=low]
 | SPLIT_WORD   rn[low_byte=low; high_byte=high; wordin=addr]
 | TRI_MUX      rn[ina=zero4; inb=low4; select=ainit; enable=init;
                    outtri=nibble1]
 | TRI_MUX      rn[ina=zero4; inb=high4; select=ainit; enable=init;
                    outtri=nibble2]
 | DRIVER8      rn[input=high; output=byte; enable=init]
 | MERGE_TRI    rn[word=drivers]
|]
hide {zero4: word4, low4: word4, high4: word4, nibble1: tri_word4,
   nibble2: tri_word4, byte: tri_word8, low: word8, high: word8}
```

ZERO4, SPLIT_BYTE, SPLIT_WORD, TRI_MUX, DRIVER8, and MERGE_TRI are given basic building blocks, and have to be defined as axioms in LCF_LSM. The implementation is depicted in figure 4.

ZERO4 always outputs the value #0000. Its specification is:

```
ZERO4 == dev {out}. {out = #0000}; ZERO4
```

This subterfuge represents a real device which is actually wired to ground. SPLIT_BYTE splits an 8-bit byte into 2 4-bit nibbles. This is achieved in the real device by wiring, but we need to be careful because we have to do the type-conversion. Its specification is:

```
SPLIT_BYTE
  ==
dev {bytein, high_nibble, low_nibble}.
{high_nibble = HIGH_NIBBLE(bytein), low_nibble  = LOW_NIBBLE(bytein)};
SPLIT_BYTE
```

SPLIT_WORD splits a word into low and high bytes. Again we need explicit type conversion. It is implemented as wiring in the real device.

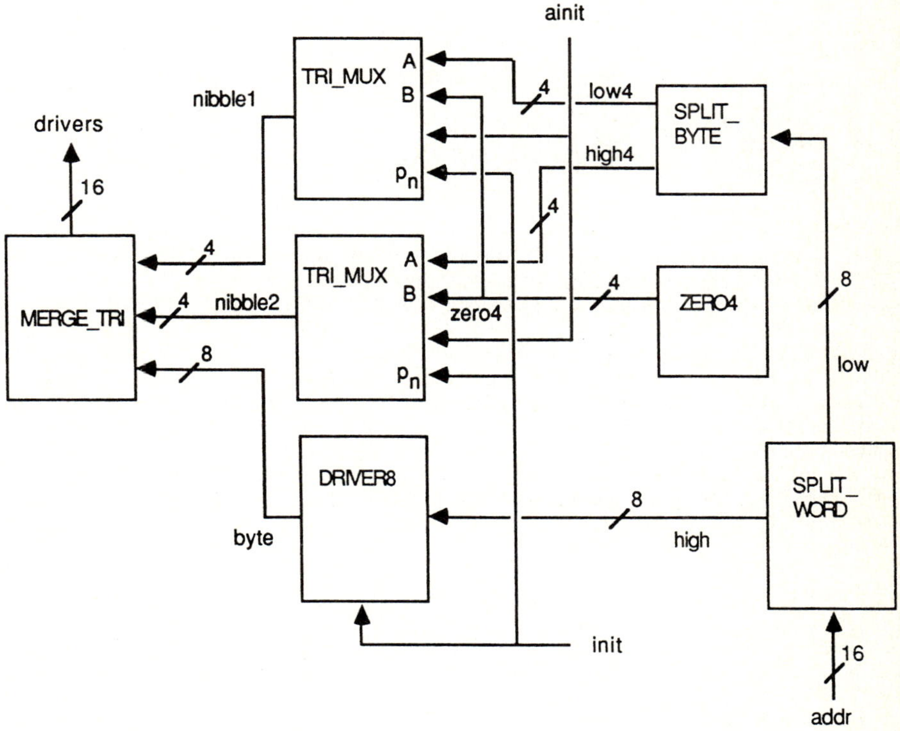

Figure 4 DRIVERS implementation

```
SPLIT_WORD
  ==
dev {wordin, high_byte, low_byte}.
{high_byte = HIGH_BYTE(wordin), low_byte  = LOW_BYTE(wordin)};
SPLIT_WORD
```

MERGE_TRI merges two 4-bit tri-state words and one tri-state byte into a 16-bit tri-state word. It is only wiring in the real device.

```
MERGE_TRI
  ==
dev {nibble1, nibble2, byte, word}.
{word = (nibble1=FLOAT4 and nibble2=FLOAT4 and byte=FLOAT8) -> FLOAT16
                   |MERGE4_4_8(nibble1, nibble2, byte)};
MERGE_TRI
```

TRI_MUX outputs 4-bit floating when disabled, otherwise selects one of two 4-bit tri-state inputs. It is implemented as a 74257 in the real device.

```
TRI_MUX
  ==
dev {ina, inb, select, enable, outtri}.
{outtri = (enable -> FLOAT4 | (select -> MK_TRI4 inb | MK_TRI4 ina))};
TRI_MUX
```

DRIVER8 is an 8-bit tri-state driver. When enabled, it outputs its input. Otherwise it outputs FLOAT8. It is a 74244 driver in the real device.

```
DRIVER8
  ==
dev {enable, input, output}.
{output = (enable -> FLOAT8 | MK_TRI8 input)};
DRIVER8
```

The DRIVERS statement of correctness is straightforward, but tedious because of type conversion boxes. Because the circuit is purely combinational, it is easy to show that DRIVERS == DRIVERS_IMP. (DRIVERS has been since proven correct automatically using a version of BARROW's VERIFY system [1].)

We stop here, having reached the level of TTL chips actually used in the design, a level that corresponds in EDICT to making use of validated library elements. We have verified all the floor plan elements in similar fashion. Working with a bipolar implementation of the design gave rise to a certain lack of elegance in the lowest level specifications, because the 8-bit functions we wanted were not available to us and we had to work with 4-bit chips. In VLSI implementations, cells at this (and higher) levels can be parameterised [14] and the resulting specifications are correspondingly clearer.

THE SECD MACHINE

The SECD machine is an architecture for executing programs written in a purely functional dialect of Lisp. The SECD machine was originally described by Landin in 1963 [13]. A version of this machine, described by Henderson [9], is the basis of our study [10].

We have designed a register-transfer level implementation of the abstract machine described by Henderson. Our design includes an register-transfer level implementation of garbage collection. We have also produced a formal specification of this implementation in the LCF_LSM hardware specification language.

Our design work began with a hardware interpreter written in a high-level programming language. This simulation software was used to investigate a possible SECD architecture down to the micro-code level. The simulation gave us insight into Landin's original SECD proposal and a complete and detailed understanding of its intended mode of operation. The simulation model of the architecture was refined precisely along the lines of the envisaged hardware decomposition. Detailed timings and critical path highlightings which emerged from simulation runs, were fed back into the next design iteration cycle. Most importantly, the simulation gave us a much clearer understanding of what the top level and floor plan specifications should contain and what they should look like.

This first-approximation freely employed high-level programming language constructs such as complex data structures and recursion. For instance, a recursive tree traversal was used in the implementation of garbage collection. The next step was to refine this high-level simulation into a register-transfer level simulation of the SECD machine. This involved eliminating the rather liberal use of high-level programming constructs and using very simple control constructs in their place. For instance, the recursive tree traversal became a non-recursive traversal. Similarly, complex data structures were replaced with a hardware oriented representation of data. In the process of refining the high-level simulation into a register-transfer level simulation, the actual architecture of the implementation gradually evolved. This was really a matter of realizing, for example, when an additional register was required in the course of translating some high-level code in the simulation into register-transfer instructions.

The effort to produce a formal specification of the implementation in parallel with development of the register-transfer level simulation was extremely worthwhile. The formal specification was used to capture the current state of the register-transfer level design. Furthermore, the formal specification was responsible for consistency in the design. For example, the formal specification enforced assumptions that the designer had 'in the back of his head' such as whether a particular signal was a 14-bit or a 32-bit value. Formal specification also encourage the designer to observe functionality in the design, for example, avoiding using a single register for two unrelated functions.

The last step was to translate the register-transfer level simulation into microcode which was then assembled into a binary image. This binary image was then used to control the lowest level simulation. This lowest level simulation actually read a microinstruction out of a (simulated) ROM, decoded it, and interpreted the control signals. The purpose of this simulation was simply to verify that the microcode has been accurately derived from the register-transfer level simulation. The simulation also provided exact (simulated) execution times.

We are currently working on a target-level specification of the SECD machine. The target-level specification of the SECD machine will closely resemble the formal description given by Henderson. Henderson describes the SECD machine in terms of a state-to-state transition for each one of the twenty-one SECD machine instructions where the state of the machine is given by the contents of the four principle registers: s, e, c and d. An sample of one of these transitions is shown below.

```
(a b.s) e (ADD.c) d  ->  (b+a.s) e c d
```

This target-level specification will be used to formally state that the implementation correctly implements the SECD machine. Proving this statement of correctness will formally verify the implementation.

The specification and verification of the SECD will involve several different types of abstractions including structural, temporal and data abstractions. We have had experience with both structural and temporal abstraction in the specification and verification of a general-purpose 16-bit microcoded computer [7, 11]. However, we expect that data abstraction will be a difficult task requiring the use of fixed-points. The required data abstraction will relate (potentially infinite) S-expressions in the target-level specification to linked lists implemented in finite RAM. The final result will be be a substantial example of hardware specification and verification.

REMARKS

Our long term aim is to build a specification based VLSI CAD system. In the short term, work on verification at Calgary has concentrated on gaining experience with large proofs. For this work, the register transfer level is the most appropriate abstraction level. More detailed mathematical models can be found in [18, 20, 21], and [19] describes initial work on an abstraction function relating timing level signals and register transfer level signals.

Implementing the elimination unit in off-the-shelf bipolar was awkward at the lowest level where we had to work with the building blocks available. These were not always what we wanted, e.g. we had to work in 4-bit nibbles when we really wanted 8-bit components. The bottom level specifications were correspondingly cluttered. We feel that this problem is due to the implementation medium, not the verification technique, for parameterised components are easy to handle in VLSI. The elimination unit proof has also taught us to be wary of attempting bottom up proofs - a major headache has been (and still is) devising a top level specification to suit the floorplan. Finally, attempting the verification of a design after its implementation was instructive. Our feelings are that implementation ad hocery begets untidy proofs and that neat specifications beget clean hardware implementations. This confirms our prejudice towards specification driven design.

ACKNOWLEDGEMENTS

The authors would like to thank Geoff Wyvill who painted the figures for us, JADE colleagues at Calgary for their help, encouragement, and general software support, and the other members of the VLSI team at Calgary for providing an stimulating milieu. We would also like to thank John Gray and Mike Gordon who have been particularly helpful not only in getting us going in verification but in sustaining us thereafter. This work is partially supported by the Natural Sciences and Engineering Research Council of Canada.

REFERENCES

[1] H.Barrow, Proving the correctness of digital hardware designs. VLSI Design, July 1984.
[2] G.Birtwistle et al., EDICT an environment for the design of integrated circuits. University of Calgary Research Report 84/155/13, 1984.
[3] J.D.Crawford, EDIF: a mechanism for the exchange of design information. CICC, 1984, 446-449.
[4] Electronic Design Interchange Format Specification. Version 1.0. 1985.
[5] M.Gordon, 1 LCF_LSM. Technical Report 41, Computer Laboratory, University of Cambridge, 1983.
[6] M.Gordon, HOL a machine oriented version of higher order logic. Draft Report, Computing Laboratory, University of Cambridge, May 1985.
[7] M.Gordon, 1 Proving a computer correct. Technical Report 42, Computer Laboratory, University of Cambridge, 1983.
[8] M.Gordon, Formal methods for hardware verification. University of Cambridge Draft Report, 1984.
[9] P.Henderson, Functional programming. Prentice Hall 1985, pages 100-150.
[10] J.Joyce, The SECD machine. University of Calgary Research Report, 1985.

[11] J.Joyce, G.Birtwistle, and M.Gordon, Proving a Computer Correct in Higher Order Logic. University of Calgary Research Report 85/208/21.
[12] P.W.Kollaritsch and N.H.E.Weste, TOPOLOGIZER: an expert system translator of transistor connectivity to symbolic cell layout. IEEE Journal of Solid State Circuits, SC-20(3), 799-804, 1985.
[13] P.Landin, On the mechanical evaluation of expressions. Computer Journal 1963.
[14] Lattice Logic, Designing with gate arrays. Edinburgh. 1981.
[15] B.Liblong, SHIFT a Structured Hierarchic Intermediate Form for VLSI Design. Masters Thesis, University of Calgary, 1984.
[16] B.Liblong and G.Birtwistle, A VLSI Design System Based Upon a High Level Intermediate Form. Canadian Conference on VLSI, Waterloo, 1983, 150-153.
[17] T-M.Lin, A hierarchical timing simulation model for digital integrated circuits and systems. PhD Thesis, Caltech 1984. Available as 5133:TR:84.
[18] G.Milne, CIRCAL and the representation of communication, concurrency, and time. Toplas, 7(2), April 1985, 270-298.
[19] T.Melham, A signal abstraction function. Draft memo, Computing Laboratory, Cambridge University. 1985.
[20] B.Moszkowski, A temporal logic for multi-level reasoning about digital hardware. IEEE Computer, February 1985, 10-19.
[21] B.Moszkowski, Executing temporal logic programs. Technical Report 55, Computing Laboratory, Cambridge University. 1984.
[22] I.M.Nixon, IF an idiomatic floorplanner. University of Edinburgh Research Report CSR-170-84, 1984.
[23] T.Patten, N.Hutchinson, and B.Unger, The flooding sink: a new approach to local area networking. University of Calgary Research Report no. 83/124/13. May 1983.
[24] M.Rem, and C.Mead, A notation for designing restoring logic circuitry. Proc Second Caltech Conference on VLSI. Pasadena 1981, 399-412.
[25] S.Trimberger, Automated performance optimisation for custom integrated circuits. PhD Thesis. Caltech 1981.

Formal Aspects of VLSI Design
G.J. Milne and P.A. Subrahmanyam (editors)

CURRENT WORK ON THE SEMANTICS OF DIGITAL SYSTEMS

Raymond T. Boute
University of Nijmegen, The Netherlands

The general context is a recast of the theory of digital systems in a functional style in order to support transformational reasoning. In particular, we will present system semantics, a uniform framework for describing system properties (structure, behaviour, performance) by means of semantic functions. These functions are suitable not only for computer interpretation but also as rules for formal reasoning. The application to digital systems covers combinational and sequential circuits, and the usage of combinators as an abstraction mechanism.

0. INTRODUCTION

Digital systems lend themselves very well to functional description. It turns out that the formal notation of a functional language is sufficiently clear and compact to replace the informal mathematical language traditionally used in the development of the theory. Hence the same language can be used for reasoning by humans (deduction of properties, theorem proving) and for interpretation by machine (verification, simulation).

Moreover, experience with this approach in teaching since 1981 has shown that many results in the theory of digital systems can be derived more conveniently in a transformational way than in the usual deductive way, and that this process often leads to new insights [3].

A simple notation based on the functional programming language SASL [19] was thereby used to describe digital systems at all "levels" (in the sense of [16]), and found adequate even for complex architectures such as the iAPX432 [11]. The only addition has been a simple type description language (used as comments in SASL), because definitions of higher-order functions tend to be unreadable without proper type information.

Apart from the intended behavioural semantics, these descriptions also allowed an intuitive structural interpretation at the combinational circuit level. The formalization of the various interpretations has led to the principle of system semantics, i.e. the description of physical systems and their properties by means of semantic functions. Its application to digital systems has thus far been elaborated up to the level of (synchronous) sequential circuits, to the extent that the conceptual problems are solved but some streamlining is still necessary.

A strong synergetic effect results from the simultaneous study of the system semantics of analog eletronic circuits [5]. The main lessons learned in this respect are:

- Compositionality is important to ensure useful formal properties (so that the semantic functions can be used for reasoning rather than just for language

description or simulation)
- Compositionality requires a careful matching between language concepts and system concepts
- The characteristics of the flow of information in a system constitute a major criterion in establishing an appropriate match.

Most of the inadequacies of current system description languages (SDLs) can be explained in terms of these observations. A recent survey [2] indicates that many SDLs are still based on imperative programming languages, as was also the case 15 years ago [17]. Imperative semantics does not allow a close match between language and system concepts, and also leads to the introduction of excessive nomenclature. Hence formal reasoning cannot be done conveniently on the basis of the semantic definitions, but requires a separate Hoare-like proof system. The corresponding proof style is deductive (based on assertions) rather than transformational.

In principle, descriptive semantics constitutes a better basis for system description. However, we have found that logic languages such as Prolog [14, 18] also have severe drawbacks, such as the proliferation of logic variables [4] and the deductive proof style. Functional languages such as Daisy [12] or muFP [15] are therefore considered more appropriate.

The collection of language concepts introduced in this paper by way of illustration constitutes a simple system description language called FUNDS (FUNctional Description of Systems). However in our treatment of system semantics, we do not aim at the complete description of a specific language, but concentrate on the relative merits of language constructs to describe system concepts. This allows more freedom regarding issues such as whether to allow variables and user-defined second-order functions, which also explains our preference for SASL over FP [1] as the semantic basis for FUNDS.

Remark. We wish to stress two additional differences between system semantics and other approaches to formal description:
- System semantics is "open" in the sense that the models are part of the formalism. This provides ready links with other formalisms that may be useful for reasoning about systems (algebra, calculus, etc.).
- The semantic domains have a well-defined type and order. Hence no theory of reflexive domains or fixpoints is required. In any case, it is felt that, in general, Constructive Type Theory is more suitable for dealing with the theoretical aspects of functional description language semantics than domain theory.

The remainder of this note is organized as follows. Section 1 describes the general context, viz. the concepts introduced for transformational reasoning about systems. Section 2 outlines the system semantics for combinational circuits, which section 3 deals with (synchronous) sequential circuits. Section 4 dicusses the definition of second-order concepts in terms of first-order ones.

1. THE CONTEXT: TRANSFORMATIONAL REASONING ABOUT SYSTEMS

Our framework for transformational reasoning has two constituents:
- A functional recast of the underlying system theory, intended for reasoning in the various domains of interpretation. In this note, we will only discuss the paradigm that is most directly relevant to digital systems, viz. representational and denotational semantics. Another formalism, the functional temporal calculus [6], is meant for reasoning in the temporal domain and unifies temporal logic, system theory and the theory of stochastic processes.
- System semantics, intended to associate meanings with formal descriptions and to reason about system properties in the notation of the formal language.

1.0 Representational and denotational semantics

Many concepts in the theory of (digital) systems can be described in terms of an abstract data type with functions of the form $f : A \to B$. The domains A and B are called domains of interest or denotation domains.

For various reasons, it may be desirable to link this description to a mapping between other domains A' and B' called representation domains. These domains may be

- more abstract, e.g. Fourier transforms of time functions,
- more concrete, e.g. binary representations of abstract values

The relationship between the two mappings is described by:

- A representation function $R : A \to A'$
- An interpretation function $I : B' \to B$
- A realization $C_f : A' \to B'$ such that $f = I \circ C_f \circ R$ ($\circ$ denotes function composition)

As discussed in a more complete report [7], R and C_f need not be functions in the proper sense but may be multiple-valued (i.e. relations, in fact). However, in case A = B we usually require that $I \circ R = id$.

Far from being an impediment, as one might be tempted to think, the explicit usage of representation and interpretation functions provides the necessary "handles" to identify rules applied in a transformational proof. Let us give just one illustrative example, viz. the proof for 2's complement addition.

Define $N_n = 0..2^n-1$ and $B = \{0,1\}$, and let $Ru_n : N_n \to B^n$ be the unsigned integer representation function with inverse $Iu_n : B^n \to N_n$. Further, let $adu : B^n \times B^n \to B^{n+1}$ be the realization of unsigned addition $+ : N_n \times N_n \to N_{n+1}$, i.e. $Iu_{n+1} \circ adu\ (Ru_n\ m, Ru_n\ m') = m + m'$ for all m, m' in N_n.

Similarly, let $Rs : Z_n \to B^n$ (with $Z_n = -2^{n-1}..2^{n-1}-1$) be defined by $Rs\ m = Ru_n\ (m \bmod 2^n)$, and $Is = Rs^{-1}$. These functions clearly describe 2's complement representation. If we define $ads = rad \circ adu$, where $rad : B^{n+1} \to B^n$ ("right adjust") omits the most significant bit and hence realizes the function mod 2^n over N_{n+1}, then the following transformational proof shows that ads realizes addition for all pairs of numbers m, m' whose sum is in the range Z_n.

$Is \circ ads\ (Rs\ m, Rs\ m')$

= {def. ads} $Is \circ rad \circ adu\ (Rs\ m, Rs\ m')$

= {def. Rs} $Is \circ rad \circ adu\ [Ru_n\ (m \bmod 2^n), Ru_n\ (m' \bmod 2^n)]$

= {adu realizes + over N_n} $Is \circ rad \circ Ru_{n+1}\ (m \bmod 2^n + m' \bmod 2^n)$

= {rad realizes mod 2^n over N_{n+1}} $Is \circ Ru_n\ [(m \bmod 2^n + m' \bmod 2^n) \bmod 2^n]$

= {property mod} $Is \circ Ru_n\ [(m + m') \bmod 2^n]$

= {def. Rs and hypothesis m + m' in Z_n} $Is \circ Rs\ (m + m')$

= {$Is = Rs^{-1}$} $m + m'$

Remark Apart from illustrating the transformational style, this proof is interesting because most textbooks treat 2's complement only by example. The only other proof found in the literature [9] is deductive, requires case analysis and is relatively long. Moreover, the application of this proof method to 1's complement has revealed an error in the overflow criterion stated in many textbooks, viz. "overflow occurs if the carry into the sign bit differs from the carry out of the sign bit". This criterion is too strict for numbers whose sum is precisely the lower bound of the representable range, as is easily verified "a posteriori" by example. Although this error is marginal and on the safe side, the fact that it was discovered in an area that is considered "well-trodden" illustrates the value of our somewhat more formal approach.

(end of remark)

The full report [7] also discusses the application to computer architecture, data transmission (error-correcting codes, data encryption) and storage.

1.1 System semantics

Given a formal language S (set of sentences), then we define a model M as an interpretation for the constants and the variables of S, together with a meaning function $m : S \to D$, where D is the domain of interpretation for the sentences.

Semantic equivalence $=_M$ between sentences is defined in the usual way, i.e. $\underline{A}s,r:S.(s =_M r \Leftrightarrow m\ s = m\ r)$. Syntactic equivalence $=_R$ is defined by taking the reflexive, symmetric and transitive closure of a set R of conversion rules defined as binary relations on S.

In accordance with the open semantics view, semantic equivalences that can be conveniently expressed by a syntactic relation are recorded as conversion rules. Hence consistency ($=_R \Rightarrow =_M$) is automatically ensured. Completeness ($=_M \Rightarrow =_R$) is largely a matter of convenience, since the system designer may have more immediate concerns with the usage of the model.

In system semantics, every property of interest (structure, behaviour, performance) is described by a (different) model. The domain of interpretation for the sentences represents the values that property may assume.

Two issues must be taken into account when applying this principle in a particular application area:

a. The level of detail provided by the language.

We say that a model M provides at least as much detail as a model M', written $M \geq M'$, if $=_M \Rightarrow =_{M'}$. It is easy to see that $M \geq M'$ iff there exists a homomorphism $h : D \to D'$ such that $m'\ s = h\ (m\ s)$ for all s in S. A model M is called a strictly initial system model if $=_M \Rightarrow =$ (syntactic equality). Such a model M is isomorphic to the syntax and satisfies $M \geq M'$ for all M'. However, in systems semantics we consider as "existing" models only those that are actually defined because they pertain to some useful property. Hence an initial system model is defined as a model M such that $M \geq M'$ for all M'. The level of detail provided by the language is by definition that of the initial system model.

For the purpose of this paper, we choose the language syntax such that the initial system model is the structural model, since this corresponds to the highest useful level of detail of interest in circuit design. This decision does not preclude the definition of semantic functions describing lower-level detail such as layout, but these functions then have the character of canonical injections into a more detailed domain of interpretation and hence are not unique (e.g. optimisation is possible).

b. Matching language concepts to system concepts

Many physical systems can be described in terms of subsystems which interact through (physically identifiable or conceptual) interfaces. By definition we consider interfaces as zeroth-order objects, subsystems as first-order objects and the patterns for putting these together as second-order concepts. This terminology envisages the following association with syntactic categories of functional languages:

order:	0	1	2
category:	variable	function	combinator

Which of these categories should actually be included in the description language requires careful consideration. The main criterion will be compositionality of the semantic definitions, i.e. the meaning of a composite must be expressible in terms of the meanings of (and no other information about) the constituents. Without this property, the semantic definitions exhibit too many complex interdependencies to be useful as rules (or "axioms") for transformational reasoning, restricting their usefulness to simulation only.

The choice of the syntax to ensure compositionality (which is most critical in the behavioral model) is determined by the characteristics of the flow of information inside the system, in particular:

- undirectionality: every (sub)system has a uniquely defined input and output interface such that input phenomena can influence output phenomena but not vice versa,
- zero loading: physical characteristics of an input interface do not influence phenomena at the output to which it is connected.

Application to digital and analog circuits has led to the following conclusions:

- For unidirectional, zero-loading systems, the generality provided by using 0th-order variables (function application being the only 2nd-order concept) does not harm compositionality.
- If the same concepts are used for unidirectional, non-zero loading systems, auxiliary semantic functions have to reflect the propagation of loading characteristics in the direction opposite the information flow, thereby reducing the degree of compositionality [5].
- For bidirectional, non-zero loading systems, the usage of 0th-order variables must be abandoned in favour of 2nd-order combinators in order to preserve compositionality. Of course, nothing prohibits the usage of (1st-order) variables as abbreviations for first-order objects.

2. SYSTEM SEMANTICS FOR COMBINATIONAL CIRCUITS

2.0 Syntax (combinational subset of FUNDS)

The free part of the abstract syntax is described in simplified BNF [20]:

function = constant | abstraction

expression = variable | application

constant = zero | one | not | and | or

variable = a | a0 | a1 | ... | z | z0 | z1 | ...

abstraction = λ(variable, ..., variable).expression

application = function(expression, ..., expression)

Restrictions are that the variables in the λ-part of an abstraction must be distinct, and that the number of expressions in an application must match the number of places associated with the function. For the constants, the number of places is 0, 0, 1, * , * in the order given (* = "any"). For an abstraction, the number of places is the number of variables in the λ-part. The syntactic categories are F (functions), E (expressions), C (constants), V (variables) and S (sentences). The sentences are defined to be the abstractions containing bound variables only.

The concrete syntax of FUNDS is inspired by SASL: we write

$\underline{\text{def}}\ f(v_0, \ldots, v_{n-1}) = e$ for $f \equiv \lambda(v_0, \ldots, v_{n-1}).e$

and $e\ \underline{\text{where}}\ (v_0 = e_0, \ldots, v_{n-1} = e_{n-1})$

for $\lambda(v_0, \ldots, v_{n-1}).e(e_0, \ldots, e_{n-1})$ (warning: unlike SASL)

Example: def select (s, x0, x1) = or(and(not s, x0), and (s, x1))

2.1 Behavioural model

The base type for the domains of interpretations is D = {0,1}. The interpretation of the constants $k : C \to D^* \to D$ (with, for an n-place constant c, $k\ c : D^n \to D$) is defined by

$k\ \text{zero} = 0, \quad k\ \text{one} = 1$

$k\ \text{not} : D \to D$ with $k\ \text{not}\ d = \overline{d}$

$k\ \text{and} : D^* \to D$ with $k\ \text{and}\ (d_0, \ldots, d_{n-1}) = d_0 \cdot \ldots \cdot d_{n-1}$

$k\ \text{or} : D^* \to D$ with $k\ \text{or}\ (d_0, \ldots, d_{n-1}) = d_0 + \ldots + d_{n-1}$

The interpretations for the variables are of type $I = V \to D$. The meaning functions for expressions, functions and sentences are:

$\text{mexp} : E \to I \to D$ with

$\text{mexp}\ v\ i = i\ v$ for variables, while for applications:

$\text{mexp}\ [\![f(e_0, \ldots, e_{n-1})]\!]\ i = \text{mfun}\ f\ i\ (\text{mexp}\ e_0\ i, \ldots, \text{mexp}\ e_{n-1}\ i)$

$\text{mfun} : F \to I \to D^* \to D$ with

$\text{mfun}\ c\ i = k\ c$ for constants, while for abstractions

$\text{mfun}\ [\![\lambda(v_0, \ldots, v_{n-1}).e]\!]\ i : D^n \to D$ with

$\text{mfun}\ [\![\lambda(v_0, \ldots, v_{n-1}).e]\!]\ i\ (d_0, \ldots, d_{n-1}) =$

$\text{mexp}\ e\ i[d_0/v_0]\ldots[d_{n-1}/v_{n-1}]$

$m : S \to D^* \to D$ with $m\ s = \text{mfun}\ s\ i$ (where i is any interpretation)

The postfix operator $[d/v] : I \to I$ is defined by $i[d/v]\ v' = (v'=v) \to d;\ i\ v'$

The conversion rules that preserve meaning are
1) The rules for α, β and η conversion of the λ-calculus.
2) Syntactic versions of the axioms for switching algebra. For obvious reasons, these are not the usual Huntington postulates [9], but the rules given in [3].

2.2 Structural model

The domain of interpretation D is the set of loop-free single output combinational circuits built from elementary gates (i.e. first-order objects). Two formulations are possible, providing an interesting contrast.

a. First-order formulation

The interpretation for the constants $k : C \to D$ associates with every constant the component suggested by that name. A typical problem in this formulation is that the number of inputs has to be derived from the number of arguments. (this can be solved in many ways, e.g. by introducing an explicit parameter). The interpretation for the variables $i : V \to D$ associates with every variable a piece of wire labelled by that name. The meaning functions for expressions, functions and sentences are:

mexp : E → I → D with

mexp v i = i v for variables, while for applications

mexp ⟦f(e_0, ..., e_{n-1})⟧ i = mfun f i together with, for j = 0..n-1,

mexp e_j i connected via its output to the j-th input of f

mfun : E → I → D with

mfun c i = k c for constants, while for abstractions

mfun ⟦λ(v_0, ..., v_{n-1}).e⟧ i = mexp e i[0/v_0]...[n-1/v_{n-1}]

(i.e. wires are relabeled by numbers)

m = mfun

Some additional rules specify which wires are considered distinct.

b. Second-order formulation

Constants are interpreted as combinators, i.e.

k : C → D^* → D with, for an n-place constant c,

k c (d_0, ..., d_{n-1}) = the gate suggested by the name c together with,

for j : 0..n-1, the circuit d_j connected via its output to the jth input of that gate.

The type of the interpretation for variables is I = V → D. The definitions of mexp and mfun are formally the same as in the behavioural model. The meaning function m : S → D is defined by m s = mfun f i where s is of the form λ(v_0, ..., v_{n-1}).e and i is defined such that it associates with v_j the jth input connector pin of the circuit.

An important observation is that the structural semantics as defined here deals with objects, whereas behavioural semantics deals with values. The notion of values versus objects has been around for some time in programming as an abstraction for referencing, and is provided explicitly in certain programming languages [13]. In structural semantics, it has the following ramifications:

- The function i : V → D maps variable names to objects. Hence different occurences of a name designate the same object. This allows to distinguish between fanout and replication, e.g.
  or(and(x, y), y) where y = or(u, v) (fanout)
  or(and(x, or(u, v)), or(u, v)) (replication)
- By virtue of the above, β-conversion does not preserve structural meaning when applied to object variables. This is remarkable because the soundness of β-conversion depends solely on the meaning functions for abstraction and application, which are the same for the behavioural model and for the second order formulation of the structural model.

In order to describe multiple-output circuits, it suffices to extend the language with a syntax for lists.

2.3 Other models

Many other useful models can be defined, such as

- The cost model, representing the component cost of a circuit.

- The sequence model, a behavioural model with base domain D = $\{0,1\}^*$, where it is assumed that in an interpretation all sequences have equal length. For an n-place constant c, k c : D^n → D, for instance:

k and $(\varepsilon, \ldots, \varepsilon) = \varepsilon$

k and $(x_0{:}d_0, \ldots, x_{n-1}{:}d_{n-1}) =$

$(x_0 \,.\, \ldots \,.\, x_{n-1}) : k$ and $(d_0, \ldots, d_{n-1})$

The meaning functions are formally the same as for the earlier behavioural model. The result corresponds to the standard model for μFP [15]. Alternatively, $D = \mathbb{N} \to \{0,1\}$, which is isomorphic to $\{0,1\}^\omega$. The definitions are straightforward, e.g.

k and $(d_0, \ldots, d_{n-1})\ j = (d_0\ j) \,.\, \ldots \,.\, (d_{n-1}\ j)$

- The hazard model, a behavioural model with base domain $D = \{0, 1, x\}$, where x stands for an indeterminate value. The definition of $k : C \to D^* \to D$ in such a way that mexp and mfun may be formally the same as in the earlier behavioural model is a simple exercise.

- The worst-case timing model with domain $D = \mathbb{R}$. For an n-place constant c, $k\ c : D^n \to D$ with $k\ c\ (d_0, \ldots, d_{n-1}) = \max(d_0, \ldots, d_{n-1}) + \text{delay}\ c$ where delay is an auxiliary function of type $C \to D$. The meaning functions mfun and mexp are the same as for the behavioural model, and $m : S \to D^* \to D$ is defined by $m\ s = \text{mfun}\ s\ i$ where $i\ v = 0$ for all variables v.

- The detailed timing model, describing behaviour in real time (taking into account actual delays) by means of the functional temporal calculus [6].

3. SYSTEM SEMANTICS FOR SEQUENTIAL CIRCUITS

3.0 Sequential functions

a. The concept of sequentiality

Consider a class of systems whose input and output signals assume values in sets I and U respectively, and whose behaviour can be described in discrete time in terms of sequences belonging to I^* and U^*. On I^* and U^* we define a binary relation $\leq$ meaning "is a prefix of", i.e. $x \leq x.y$ (. means concatenation).

A (partial) function $f : I^* \to U^*$ is called sequential if $x \leq y \Rightarrow fx \leq fy$ (for all x, y in I^* for which f is defined)

It is easy to prove that $f : I^* \to U^*$ is sequential iff there exists a function $f' : I^* \to U^*$ such that $f(x.i) = (fx).f'(x.i)$ for all x in I^* and all i in I (where I is identified with I^1). This function f' is called the derivative of f. It is clearly unique if we also require that $f'\varepsilon = f\varepsilon$ and that $f'x$ is defined iff fx is defined.

b. The concept of state

Given a sequential function f and its derivative f', we define a Curry-ed infix operator $\uparrow$

$\uparrow : I^* \to I^* \to U^*$ with $x{\uparrow}y = f'(x.y)$

We call $x{\uparrow}$ (of type $I^* \to U^*$) the state after x, and $Q = \{\, x{\uparrow} \mid x : I^*\}$ is the state set. The notation ${\uparrow}x$ stand for $\varepsilon{\uparrow}x$, where ε is the empty sequence.

It is easy to show that $\uparrow$ has the following properties:

i) $\uparrow$ completely defines f (insofar as f is defined)

ii) $x{\uparrow}y.z = x.y{\uparrow}z$ for all x, y, z in I^*

The set Q is isomorphic to the set $I^*/\equiv$ of Nerode equivalence classes in the sense that $x{\uparrow} = y{\uparrow}$ iff $[x] = [y]$. However, the formulation of the concept of state as a higher-order function has more useful transformational properties, as illustrated next.

c. Example [8]

This example is a transformational recast of the classical deductive proof showing that $\{0^k 1^k \mid k : \mathbb{N}\}$ is not regular.

Given the following specification of a sequential function f in terms of its derivative f' :

$$f' : \{0,1\}^* \to \{0,1\} \text{ with } f'x = 1 \Leftrightarrow \underline{E}k{:}\mathbb{N}.(x = 0^k 1^k).$$

Let $Q = \{ x\uparrow \mid x : I^* \}$ and $Q_0 = \{ 0^k\uparrow \mid k{:}\mathbb{N} \}$. Clearly Q_0 is a subset of Q. If Q is finite, say $|Q| = n$, then $|Q_0| \leq n$ and hence $\underline{E}i,j{:}N.(0 \leq i < j \leq n$ and $0^i\uparrow = 0^j\uparrow$). For these values of i and j,

$$f'(0^j 1^j) = \{\text{definition } \uparrow\}\ 0^j\uparrow 1^j$$
$$= \{0^i\uparrow = 0^j\uparrow\}\ 0^i\uparrow 1^j$$
$$= \{\text{definition } \uparrow\}\ f'(0^i 1^j)$$

Since $i \neq j$, this is incompatible with the specification.

3.1 Semantics of sequential machines

a. Representational and denotational semantics

The domain of interest is the set of sequential functions $F \subset I^* \to U^*$. For the representation domain we choose the set of generalized sequential machines (in the sense of Ginsburg):

$$M(I, U, Q) = (Q \to I \to U^*) \times (Q \to I \to Q) \times U^* \times Q$$

which must be seen as a type parametrised by I, U, Q.

A typical element (out, next, prel, start) : M is interpreted intuitively as follows: output function, next-state function, prelude (output produced at startup) and starting state. Formally, this interpretation is defined by the following behaviour function.

```
def || B : M(I, U, Q) → I* → U* with
        B m x = prel ++ (G start x)
              where || G : Q → I* → U* with
                       G q () = ()
                       G q (i:x) = (out q i) ++ (G (next q i) x)
                       (out, next, prel, start) = m
```

Conversely, we define the following representation function:

```
def || R : (I* → U*) → M(I, U, I* → U*) with
        R f = (out, next, prel, start)
            where out q i = q (i,)
                  next q i x = q (i : x)
                  prel = f ε
                  start = f'
                  f' = derivative f
```

It is easy to prove that R and B are each other's inverse.

A state machine is a machine in which U = Q, out = next and prel = ε. Hence for state machines MS(I, U, Q) = (Q → I → Q) x Q

Remarks
- As written, B and R are executable SASL functions.
- The definition of B is very convenient for deriving various results about sequential machines, in particular those pertaining to state and machine equivalence and to state reduction [8].
- The "behaviours" in the sense of Gordon [10] can be defined directly in terms of M(I, U, Q), whereas the converse is somewhat more difficult. In this sense, our definitions are more symmetrical.

b. System semantics

The following is a formalization of a notation we have been using since 1981 for the description of the synchronous sequential circuits.

Syntax
The combinational subset of FUNDS is extended by the free part:

```
constant = memorycell initialvalue
     memorycell   = dff | tff | jkff | rsff | ...
     initialvalue = 0 | 1
application =  expression feedback (variable = expression, ..., variable =
                                         expression)
```

together with the following restrictions:
- The constants derived from dff and tff are one-place, the others are two-place.
- Syntactic restrictions to exclude closed loops (in the structural model). Rather than elaborating a set of necessary and sufficient conditions for arbitrary (and hence unstructured) combinations, we impose the following sufficient condition which corresponds to the standard (structured) form under which any synchronous circuit can be written: in an application of the form

  $$e \text{ feedback } (v_0 = e_0, \ldots, v_{n-1} = e_{n-1})$$

  every e_j must be of the form $g_j\ b_j\ e'_j$ or $g_j\ b_j\ (e'_j, e''_j)$, where g_j is a memory cell.

In an application of the above form, all occurrences of $v_0, \ldots, v_{n-1}$ are by definition bound, in contrast with the "where" construct.

Example: divider by 4

```
def div4 in = and(q1, not q0)
              feedback q0 = dff 0 select(in, q0, not q1)
                       q1 = dff 0 select(in, q1, q0)
```

Structural model
For the newly introduced constants of the form c = g b,

$$k\ [\![g\ b]\!] : D^n \to D \quad (n = 1 \text{ or } 2)$$

$k\ [\![g\ b]\!]\ (d_0, d_1)$ = the memory cell of the kind suggested by g together with the excitation circuits d_0, d_1

For the feedback application,

$$m\ [\![e \text{ feedback } v_0 = e_0, \ldots, v_{n-1} = e_{n-1}]\!]\ i = m\ e\ i'$$
$$\text{where } i'\ v = (v = v_j) \to m\ e_j\ i';\ i\ v$$

Since this interpretation pertains to physical rather than mathematical objects, the significance of the recursive part of the definition, viz. $i'\ v_j = m\ e_j\ i'$, must be explained in words: the output of $m\ e_j\ i'$ is connected as feedback to the input of the same circuit corresponding to v_j.

Behavioural model

For the domain of definition, one can choose between $D = \{0,1\}^*$ and $D = \mathbb{N} \to \{0,1\}$. For a memory cell g we introduce an auxiliary function k' defined as follows: $k'\ g : \{0,1\} \to D^n \to D$ $(n = 1, 2)$ with, for the case $D = \{0,1\}^*$,

- $k'\ dff\ q\ d = q{:}d$

- $k'\ tff\ q\ () = q,$
  $k'\ tff\ q\ (x{:}d) = q : k'\ tff\ (x = 0 \to q\ ;\ \overline{q})\ d$

- $k'\ jkff\ q\ () = q,$
  $k'\ jkff\ q\ (x{:}d,\ x'{:}d') = q : k'\ jkff\ (q = 0 \to x\ ;\ \overline{x}')\ (d, d')$

If $D = \mathbb{N} \to \{0,1\}$ these definitions become

- $k'\ dff\ q\ d\ n = (n = 0) \to q\ ;\ d(n - 1)$

- $k'\ tff\ q\ d\ n = (n = 0) \to q\ ;\ d(n - 1) = 0 \to p\ ;\ \overline{p}$
  where $p = k'\ tff\ q\ d\ (n - 1)$

and so on. Other styles of definition are readily apparent. On the basis of these auxiliary definitions, we define

$k\ [\![g\ b]\!] = k'\ g\ (k\ b)$ with $k\ 0 = 0$ and $k\ 1 = 1$

The definition of the meaning function for the feedback application is formally the same as for the structural model. The reader should satisfy himself that the recursive part in the semantic definition $(i'\ v_j = m\ e_j\ i')$ presents no problems because $e_j = g_j\ b_j\ e'_j$.

4. SECOND-ORDER CONSTRUCTS

For combinational circuits, the usage of zeroth-order variables in conjunction with abstraction and application is found to be the most appropriate way for describing loop-free interconnection patterns, because:

- it completely satisfies the principle of compositionality,
- it does not cause proliferation of variables since these are used only to express fanout (or sometimes to enhance clarity),
- it reflects the data paths more clearly than the selectors of FP-like languages.

Experience with FP and with the design of digital description languages based on FP [5] indicates that selectors tend to make descriptions difficult to read, even after extensive practice. This is perhaps not surprising in view of the fact that mathematics texts from antiquity contain phrases such as "the square root taken from the square of the second quantity diminished with four times the product of the first and the third quantity", whereas the introduction of variables in a later period is considered as one of the major breakthroughs in mathematics. Nevertheless, variables should be used judiciously.

On the other hand, second-order concepts such as combinators or combining forms constitute a valuable complement to the lower-order concepts, because they can serve as abbreviations for complex compositions and hence contribute to a structured description style. However, rather than introducing combining forms as the basic constructs of the language, thereby excluding (as in FP) both user-defined combinators and lower-order concepts, we allow all combinators to be user-defined precisely in terms of the lower-order concepts. In this fashion, the finer grain of the description can be exposed or hidden as necessary.

In order to capture the meaning of such combinators simultaneously for all interpretations, we define them syntactically by means of rewriting rules. Their elaboration can be seen as macro expansion which takes place only when actual "arguments" (tuples, functions) are supplied. This allows us to bypass the need for separate semantic definitions.

In the following, we assume that the language FUNDS has been extended with tuples and with functions having multiple "result" places. Such an extension is necessary for multiple-output circuits anyway and is fairly straightforward. It is often convenient to introduce a syntactic prefixing operator (:) and formal recursion on tuples (elaborated by unfolding) into the macro definitions. Furthermore, as long as no type system is included in the language, we tacitly assume that the number of formal arguments (input, output places) of the composed descriptions match.

4.0 Examples of combinators for circuit composition

In order to simplify the syntactic expansion of the expressions, we give context-dependent rewriting rules, which amounts to applying appropriate conversion rules in advance.

Series connection (symbol $\circ$) of two circuits f and g:
   isolated: $f \circ g = \lambda v.f(g\ v)$
   in context: $f \circ g\ e = f(g\ e)$

Parallel connection (symbol [ ]) of two circuits f and g:
   isolated: $[f, g] = \lambda v.(fv, gv)$
   in context: $[f, g]\ e = (fv, gv)$ where $v = e$

Replication (symbol @) of a circuit f ($m \geq 1$):
   for an m-output expression e
      $@f\ e = @f\ (v_0, \ldots, v_{m-1})$ where $(v_0, \ldots, v_{m-1}) = e$
   for tuples of expressions
      $@f\ () = ()$
      $@f\ (e : rest) = f\ e : @f\ rest$

Right insertion (symbol /R) of a 2-input circuit f ($m \geq 1$):
   for an m-output expression e
      $/Rf\ e = /Rf\ (v_0, \ldots, v_{m-1})$ where $(v_0, \ldots, v_{m-1}) = e$
   for tuples of expressions
      $/Rf\ (e) = e$
      $/Rf\ (e : rest) = f\ (e, /Rf\ rest)$

Left insertion and other interesting combinators equivalent to the FP combining forms can be defined similarly. Remark that the definition of combinators in terms of more elementary constructs allows to tie up a few loose ends in the definition of μFP [15]. Indeed, μFP defines combining forms in terms of sequences, but in reality they are applied to (multiple-output) functions. The ability to introduce variables via the rewriting rules overcomes this discrepancy.
We conclude with an expansion example. If g is a function with m output places, then
   $(/Rf) \circ g$

      = {expansion $\circ$} $\lambda x.(/Rf\ (gx))$

      = {expansion /Rf for m-output expression gx}

        $\lambda x.(/Rf\ (x_0, \ldots, x_{m-1})$ where $(x_0, \ldots, x_{m-1}) = gx)$

      = {expansion /Rf for m-tuple $(x_0, \ldots, x_{m-1})$}

        $\lambda x.(f(x_0, f(x_1, \ldots, f(x_{m-2}, x_{m-1}) \ldots))$ where $(x_0, \ldots, x_{m-1}) = gx)$

4.1 Example of a combinator for sequential circuits

The μ-combinator of μFP is easily described in FUNDS as follows:

μf = λx.[u feedback (u,y) = (u, dff 0 v) where (u,v) = f(x,y)]

(Parentheses are omitted subject to the convention that "where" binds more closely than "feedback". Hence u is not fed back).

Remark that the finer grain of FUNDS makes it possible to introduce different kinds of flipflops.

CONCLUSION

This note provides a representative overview of the ways in which the transformational style of reasoning is being introduced in the theory of digital systems, and of the present status of the work towards formalizing the supporting notation. Additional streamlining is required for the second-order system semantics of sequential circuits. A promising approach appears to consist in formalizing and extending some of the concepts of μFP and in introducing a formal type definition language into the SDL itself (i.e. not just for usage at the meta-level). Experience with our notation for describing complex computer architectures indicates the desirability of such a type language for other purposes as well.

REFERENCES

[1] Backus, J., "Can programming be liberated from the von Neumann style? - A functional style and its algebra of programs", Comm. ACM 21, 8 (Aug. 1978) 613-641.

[2] Barbacci, M., and Uehara, T., guest editors, IEEE Computer 18, 2 (issue on hardware description languages) (Feb. 1985) 42-102.

[3] Boute, R.T., Inleiding tot de structuur van digitale systemen, Course notes M1, Computer Science Department, University of Nijmegen (Aug. 1983).

[4] Boute, R.T., Programs versus descriptions, Proc. SEAS Spring meeting (April 1984) 451-467.

[5] Boute, R.T., System semantics applied to digital and analog circuits, Report no. 56, Computer Science Department, University of Nijmegen (Aug. 1984).

[6] Boute, R.T., Functional temporal calculus versus temporal logic: an engineering approach, Report no. 59, Computer Science Department, University of Nijmegen (March 1985).

[7] Boute, R.T., Representational and denotational semantics of digital systems, Report no. 61, Computer Science Department, University of Nijmegen (in preparation).

[8] Boute, Functional description of digital systems, In: Giloi, W.K. and Shriver, B.D. (eds.), Methodologies for computer system design (North-Holland 1985).

[9] Friedman, A.D., Logic design of digital systems, Computer Science Press (Woodland Hills, 1975).

[10] Gordon, M.J.C., The denotational semantics of sequential machines, Information processing letters 10, 1 (Feb.1980) 1-3.

[11] Intel Corp., iAPX432 General Data Processor Architecture Reference Manual, Intel Corp., order number 171860-004. Santa Clara, Ca. (1983).

[12] Johnson, S.D., Synthesis of digital designs from recursion equations, The MIT Press, (Cambridge, Mass. 1984).

[13] Maclennan, B.J., Abstraction in the Intel iAPX432 Prototype Systems Inplementation Language, ACM Sigplan 18, 12 (Dec. 1983) 86-95.

[14] Maruyama, F., and Fujita, M., Hardware verification, IEEE Computer 18. 2 (Feb. 1985) 22-32.

[15] Sheeran, M., muFP: a language for VLSI design, 1984 ACM Symposium on LISP and functional programming (Aug. 1984) 104-112.

[16] Siewiorek, D.P., and Bell, C.G., Computer structures: principles and examples McGraw-Hill, New York (1981).

[17] Stabler, E.P., "System description languages", IEEE Trans. Comp. C-19, 12 (Dec. 1970) 1160-1173.

[18] Suzuki, N., Concurrent Prolog as efficient VLSI design language, IEEE Computer 18, 2 (Feb. 1985) 33-40.

[19] Turner, D.A., SASL Reference Manual (revised version), University of Kent (1979).

[20] Wirth, N., What can we do about the unnecessary diversity of notation for syntactic definitions?, Comm. ACM 20, 11 (Nov. 1977) 822-823.

Formal Aspects of VLSI Design
G.J. Milne and P.A. Subrahmanyam (editors)

Automatic Circuit Verification Using Temporal Logic: Two New Examples

Michael C. Browne
Edmund M. Clarke
David L. Dill

Carnegie-Mellon University
Pittsburgh Pennsylvania 15213

1. Introduction

Temporal logic is a formal system for reasoning about the occurrence of events in time. It belongs to the class of modal logics, originally developed by philosophers in an attempt to classify logical propositions according to their "possibility" or "impossibility". In the case of temporal logic special operators are introduced that describe how the truth values of assertions vary with time. A typical operator is **G** *f* which is true now if *f* is true at all future moments (*f* is Globally true). As an example of how temporal logic might be used, consider the assertion that two events S_1 and S_2 do not occur simultaneously. This is naturally expressed by the formula $\mathbf{G}(\neg S_1 \vee \neg S_2)$.

Pneuli [14] was apparently the first to realize that temporal logic might be useful for reasoning about concurrent programs. Later, Bochmann [4] and Owicki and Malachi [13] showed how his ideas could be extended to handle sequential circuits. Although these researchers contributed significantly toward developing an adequate notation for specifying properties of circuits, the problem of showing that a circuit actually met its specifications still required a tedious hand-constructed proof.

In previous papers ([6], [7]) we have described an automatic, temporal logic based verifier for finite state concurrent systems. Our verifier uses a simple and efficient algorithm, called a *model checker*, to determine the truth of a temporal formula relative to a state transition graph. If the formula is not true, the model checker will provide a counterexample if possible. Unlike most verifiers, our algorithm has time complexity linear in both the size of the specification and the size of the state-transition graph. Experimental results show that the verifier can check temporal properties of state transition graphs at a rate of roughly a hundred states a second. Moreover, since the algorithm is linear in the size of the state graph we can expect the same rate for graphs with several thousand states that we get for graphs with several hundred states.

Sequential circuit verification is a natural application for the verifier. In ([1], [11], [8]) we have described how the verifier might be used to debug such circuits. The basic idea is to extract a state transition graph from some representation of the circuit and then use the model checker. In this paper we provide further evidence for the usefulness of our approach by describing enhancements to the basic verifier that automate the extraction of state transition graphs from circuits. We discuss two different techniques. The first approach involves extracting the state graph directly from a wire-list description of the circuit and is suitable for asynchronous circuits. The second obtains the state diagram by compilation from an HDL specification of the original circuit. Although these approaches are quite different, we believe that there are situations in which each is useful.

2. The Logic and the Model Checker

The logic that we use to specify circuits is a propositional temporal logic of branching time, called CTL (Computation Tree Logic). It is similar to those described in [3], [7], and [12]. The syntax that we use for CTL in this paper is given below. We assume the existence of an underlying set of atomic propositions *AP* for denoting events.

- Any atomic proposition $P \in AP$ is a CTL formula.
- If *f* and *g* are CTL formulas, then $\neg f$, $f \wedge g$, $\mathbf{G} f$, $\mathbf{F} f$, $f\, \mathbf{U}\, g$, and $f\, \mathbf{u}\, g$ are CTL formulas.

The symbols $\neg$ and $\wedge$ have their usual meanings. In addition to these formulas, we use the abbreviations $f \vee g$ and $f \rightarrow g$ for $\neg(\neg f \wedge \neg g)$ and $\neg(f \wedge \neg g)$, respectively. Intuitively, $\mathbf{G} f$ means that *f* is true in every state (Globally true), and $\mathbf{F} f$ means that *f* is eventually true somewhere along every sequence of states (true at some Future time). There are two versions of the *until* operator: the *strong until* "**U**" and the *weak until* "**u**". The *strong until* $f\, \mathbf{U}\, g$ requires that *g* must eventually be true and that *f* always hold until *g* is true. The *weak until* $f\, \mathbf{u}\, g$ also requires that *f* always be true until *g* becomes true; however, it is permissible for *g* never to be true if *f* remains true forever.

The semantics of CTL formulas is defined with respect to a labeled state transition graph. A CTL *structure* or *model* is a state transition graph consisting of:

- A finite set of states S;
- A function $L: S \rightarrow \mathcal{P}(AP)$, which labels each state with a set of atomic propositions;
- A transition relation $R \subseteq S \times S$;
- A start state, s_0.

We require that every state have at least one successor. A *path* from $s_i \in S$ is an infinite sequence of states that starts with s_i and in which R holds between every state and its successor.

Below we formally define what it means for a formula f to be *true in a state* s_i. This is written $s_i \models f$.

- $s_i \models P$ iff $P \in L(s_i)$.
- $s_i \models f \wedge g$ iff $s_i \models f$ and $s_i \models g$.
- $s_i \models \neg f$ iff it is not the case that $s_i \models f$.
- $s_i \models \mathbf{G} f$ iff for every state s_j on every path from s_i, $s_j \models f$.
- $s_i \models \mathbf{F} f$ iff there exists a state s_j on every path from s_i such that $s_j \models f$.
- $s_i \models f \, \mathbf{U} \, g$ iff for every path from s_i there exists a state s_k such that $s_k \models g$ and for all states s_j preceding s_k on the path, $s_j \models f$.
- $s_i \models f \, \mathbf{u} \, g$ iff for every path from s_i either
  1. there exists a state s_k such that $s_k \models g$ and for all states s_j preceding s_k on the path, $s_j \models f$, or
  2. for all states s_j on the path $s_j \models f$.

A formula is true in a state graph iff the formula is true in s_0. Figure 2-1 shows some examples of CTL formulas and state graphs in which they are true.

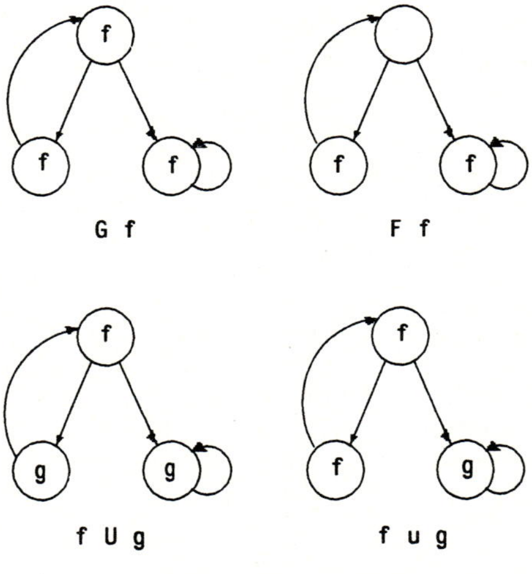

Figure 2-1: CTL Formulas and State Graphs in Which They Are True

There is a program called EMC ("Extended Model Checker") that verifies the truth of a formula in a model using these definitions. The algorithm processes a formula bottom up, checking the shortest subformulas before checking the subformulas that contain them. When it checks each subformula f, it labels every state in the graph with f if f is true in the state. Thus, when EMC is processing a formula it can treat the subformulas as atomic propositions. Processing any of the modal or propositional connectives requires at most a single depth-first traversal of the state graph, so the time to check a formula is proportional to the size of the state graph and size of the formula.

There is one very practical feature of the model checker that should be described: the counterexample facility. When the model checker determines that a formula is false, it will attempt to find a path in the graph which demonstrates that the negation of the formula is true. For instance, if the formula has the form **G** f, our system will produce a path to a state in which $\neg f$ holds. This feature is quite useful for debugging.

EMC is written in C and runs on a VAX 11/780 under Unix.

3. Verifying Asynchronous Circuits

The difficulty of designing correct asynchronous sequential circuits is well-known. Unlike synchronous circuits, in which concurrent activities proceed in lock-step, asynchronous circuits can have a large number of alternative execution sequences, each resulting from a different set of delays in the circuit elements. The circuit must ultimately produce correct results for any of these executions.

To be certain that an asynchronous design is correct, the designer must make delay assumptions that are guaranteed to hold for all instances of the circuit. One common delay model assumes that there is an arbitrary finite delay on every gate output. We term designs relying on this assumption *speed-independent.* This is a very conservative model, in the sense that any circuit that appears to be correct will work under a wide variety of realizations. However, it is often quite difficult to design circuits that are truly speed-independent. When it is possible, the resulting circuits are often unnecessarily complex and expensive, since "any reasonable implementation" would satisfy stronger delay assumptions.

We believe that the design of correct and practical asynchronous circuits could be greatly facilitated by a verification tool to automate the tedious analysis of circuit executions under varying delays. This tool should be able to use the designer's assumptions about circuit delays to avoid considering unreasonable circuit executions.

In the remainder of this section we describe in more detail the speed-independent circuit delay model and the algorithm for extracting a state graph from a circuit. We then discuss the delay assumptions and how they are incorporated into the state-graph construction algorithm. Finally, as an example, we describe, specify, and verify a "real" asynchronous circuit: a patented queue element.

3.1. Speed-Independent Circuits

The structural description of a circuit consists of a set of *elements*, which have *inputs* and *outputs*, and *nodes*, which are used to interconnect element inputs and outputs. Every element input and output must be connected to a node. Every node must be connected to *exactly one* element output and can be connected to any number of inputs.

Elements represent circuit components, which compute values from their inputs and, after some delay, supply these values on their outputs. Circuit elements are described using flow tables. Each element has a set of possible internal states. There is a *transition function* which gives a set of successor states, given a current state and values for the inputs to the element. There is also a *stability predicate* which applies to a state and an input assignment, and returns "true" if the element can stay in that state/input configuration forever. A "false" value for the stability predicate indicates that the element can stay in the configuration arbitrarily long, but must change state in finite time. Finally, there is an *output* function which maps the current internal state of the element to boolean values for its outputs.

A flow table diagram for a two-input NOR gate appears in figure 3-1. The flow table consists of two parts: on the left is the transition/stability table, with states labeling rows and input assignments labeling columns. The transition function is represented by arrows and the stable states are circled. The effect of this representation is to model a gate as an instantaneous boolean function with an *inertial* delay on its output (i.e., the case where the inputs of an inertial delay change twice before the output changes, the first change is ignored).

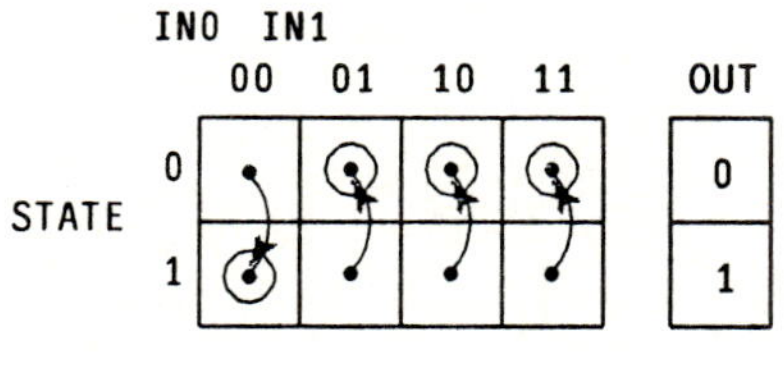

Figure 3-1: Flow Table for 2-input NOR Gate

The generality of the flow table model is useful when elements other than boolean gates are modeled. For example, many asynchronous circuits (but not our example) are designed using primitive elements other than gates (for example, flip-flops). In our example, more complex flow tables are used to model the external environment of the circuit.

Using this element model, a structural circuit description can be converted into a state graph. Each state in the global state graph is a tuple of element states, one state from each element. The global state defines a set of values for the nodes, by taking the combined effect of the output functions for each of the element states in the global state tuple (remember that the output for each element depends only on its internal state). This gives a value for every node, since every node is connected to exactly one element output. The global state is labeled with the names of the nodes having "1" values in the state, for later use by EMC.

Given a global state and its node values, the successors of the state can be found by applying the transition functions of the individual elements. We choose to use an "interleaved" model of concurrency, in which each successor of the global state corresponds to a state change in exactly one of the elements. The global state graph for a correct circuit is often of quite reasonable size, since only the states reachable from a user-specified initial state need to be represented. This is easily accomplished by constructing the graph recursively in a depth-first manner, starting with the initial state. States are stored in a table as they are created; whenever an old state is encountered, the recursion stops.

A more subtle aspect of the element model and state-graph construction is the need for stability predicates. In order to verify "liveness" properties of a circuit, which assert that it makes progress towards some state, it is necessary to avoid checking paths in the state graph in which an element remains forever in an unstable configuration. The problem is solved using the *fairness constraint* facility of EMC. A name is invented for each state in each circuit element. Every global state is labeled with the names for the unstable element states in it, as determined by the stability predicates. In addition to the state graph, EMC is supplied with a set of fairness constraints requiring each of the unstable state names to be *false* infinitely often. In this way, CTL formulas are checked only over those paths in which no element stays in an unstable state infinitely long.

This approach has been used to discover a timing problem in a published speed-independent arbiter design [11] and to verify a corrected version of the design.

3.2. Almost Speed-Independent Circuits

The circuit model and state-graph construction described so far are sufficient to verify truly speed-independent circuits. However, many asynchronous circuits that are not strictly speed-independent work perfectly well in practice. Such circuits are often defended on the grounds that they would work properly, assuming that the delay in one element is always less than the delay in another. We can formalize a delay model which includes assumptions of this type. We call circuits designed under this delay model *almost speed-independent.*

The almost speed-independent state graph of a circuit is constructed by associating with each global state a set of relationships (called *delay constraints*) of the form $x < y$, where x and y are the names of elements that are unstable in that global state. If $x < y$ is associated with a global state, it means that "x will change state before y does." In this case, the global state will have a successor representing a change in x but no successor for a change in y. Under a speed-independent model, either x or y could change.

Two states are considered to be equivalent if their element tuples are the same *and* if their delay constraints are the same. (This is important for the table lookup in the state-graph construction.) The resulting state graph is similar to the speed-independent state graph, except paths that fail to satisfy the user-supplied delay constraints have been removed. This model and algorithm are described in greater detail in another paper [10].

3.3. Example: An Asychronous Queue Element

As an example, we verify an asynchronous queue cell patented by Cogar in 1965 [9]. The suggested application for this circuit is as a buffer for reading data from recordings. A schematic appears in figure 3-2. The original circuit diagram was depicted as a collection of diodes and transistors; we have coallesced these into NOR gates.

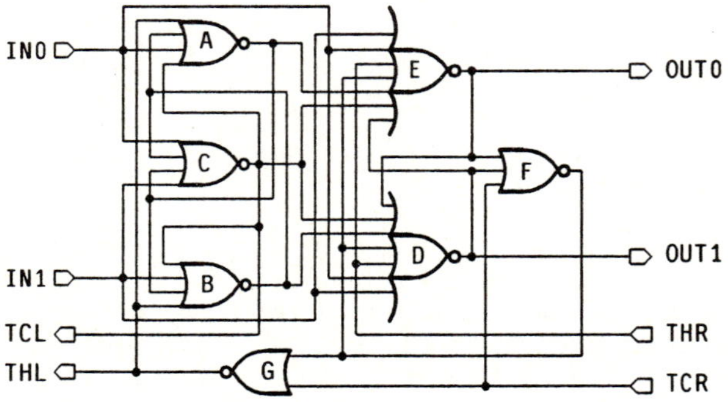

Figure 3-2: Logic Diagram for Cogar Queue Element.

The queue cells are intended to be chained together with a data source at one end and a data sink at the other. When a queue cell is empty and data is presented at its input, it reads the data and stores it internally, then makes it available at its output. In the discussion below, we consider a queue cell in the middle of a chain of queue cells in which data flows from left to right. The left neighbor of the cell can either be another queue cell or a data source; the right neighbor, either a queue cell or data sink.

The data is encoded on two wires, a "one" wire and a "zero" wire. A 1 signal is encoded as a high value on the "one" wire and a low value on the "zero" wire. Similarly a 0 is represented by a low value on "one" and a high value on "zero". "No data" is represented by a low signal on both wires, and it is never the case that both wires are high. Each queue cell has a pair of input data wires entering from the left, and output data wires exiting from the right.

In addition to the data wires, there are two control wires going from right to left. Each cell has a "transfer and hold" (TH) wire and a "transfer and clear" (TC) wire entering from the right (THR and TCR) and exiting to the left (THL and TCL). A high value on TH indicates that the cell generating it is about to become empty. A high value on TC signals that the cell is actually empty and may receive data from the cell on its left when TH goes low again.

The exact sequence of signals is shown in figure 3-3. For the interface between the cell and its left neighbor, DATA is the logical OR of IN0 and IN1, TH is THL, and TC is TCL. For the right neighbor, DATA is OUT0 v OUT1, TH is THR, and TC is TCR. The heavy arrows indicate that if the first event occurs, the second must also occur – and furthermore, the events must occur in order shown. The light arrows specify that the second event may or may not occur, but if it does occur it *must* happen after the first event. The two light arrows represent the cases where the left cell provides no data, and where the right cell stops reading data.

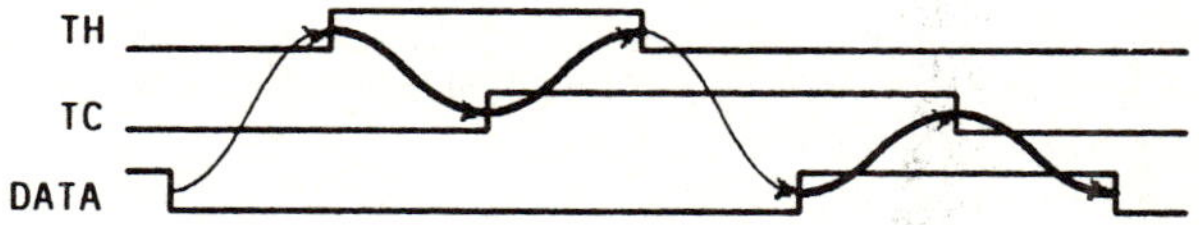

Figure 3-3: Timing Diagram for Cogar Queue Element.

This timing diagram can be expressed in CTL. Each transition can be uniquely characterized by the values of the signals immediately after it occurs. The arrows impose a total ordering on the states, which can be specified in CTL by saying, in essence, "if the signals are in the state immediately after the tail of an arrow, they must stay in that state until they take on the values of the state after the head of the arrow". The light arrows are represented by formulas using **u** and the solid arrows by **U**. For example, the first two arrows in the diagram are expressed by:

$$\mathbf{G}((\neg TH \wedge \neg TC \wedge \neg DATA) \rightarrow (\neg TH \wedge \neg TC \wedge \neg DATA)\ \mathbf{u}\ (TH \wedge \neg TC \wedge \neg DATA)), \quad \text{and}$$

$$\mathbf{G}((TH \wedge \neg TC \wedge \neg DATA) \rightarrow (TH \wedge \neg TC \wedge \neg DATA)\ \mathbf{U}\ (TH \wedge TC \wedge \neg DATA))$$

A queue cell can store one of three "values": 0, 1 and *empty*. If a cell is empty, its left neighbor will detect this by TCL being high. If the neighbor has a stored value, it will put the data on IN0 and IN1 of the cell. The cell then stores this value and lowers TCL. When the left neighbor sees TCL go low, it becomes empty and lowers both IN0 and IN1.

Even after it has stored data, the cell continues to supply a "no data" signal on OUT0 and OUT1 until its right neighbor has lowered THR and raised TCR. It then puts its stored value on OUT0 and OUT1. After the right neighbor has stored this, it will lower TCR, causing the cell to clear itself (and eventually remove the data signals). During the time when the cell is clearing itself, THL goes high to inhibit the transmission of data by the left neighbor since TCL may rise before the cell has stabilized in the "empty" state. THL falls after TCL has risen, to indicate that the cell is both empty and ready to receive new input.

The gate-level implementation of the queue cell appears in figure 3-2. The gates A, B, and C together store the data internal to the cell. When the circuit is stable, only one of them may be high at a time. If the high signal is A, a 1 is stored, if it is B, a 0 is stored, and if it is C, the cell is empty. The gates are cross-wired to ensure that only one is high at a time. Additionally, IN0 is an input to gates A and C and IN1 is an input to B and C, so the high input wires inhibit the gates that should not be high (thus, A and B are both inhibited when there is no data). The output of C becomes TCL, which indicates to the left neighbor that the gate is empty.

The gates D and E provide the signals OUT1 and OUT0 to the right neighbor. Once again, only one of these gates should be on at one time. The signals IN0 and IN1 are inputs to these gates, so they are both inhibited when the cell is storing data from its left neighbor. They are also cross-wired so that only one is high at a time. The output of gate C goes to both gates, so the outputs are low when the cell is empty. Also, the output of A goes to E so that OUT0 is low when a 1 is stored; similarly, the output of B goes to D. THR is an input to both gates so that output is suppressed when the right neighbor is being cleared. Finally, the output of gate F is fed back to both of the gates. This signal is high only when both D and E are low and when TCR is low. In the case where the cell is empty, its right neighbor is full, and the cell stores an input from its left neighbor, the F signal serves to lock the cell outputs low until TCR goes high (signaling that the right neighbor has become empty).

Gate G provides the signal to clear the cell when its right neighbor has stored the cell output. Suppose that the output of gate F is high, due to the circumstances described in the last sentence of the preceding paragraph. When the right neighbor raises TCR, F will eventually go low. This allows the cell to transmit its output to the right neighbor (assuming it is not empty). When the right neighbor has stored the value, TCR will go low. This raises the output of G, which both clears the cell (by lowering A and B), and raises TCL, which turns off output of the left neighbor until the cell has stabilized.

We have not considered a number of other possible sequences of events. These might arise due to differences of speeds of the various gates when input changes are presented simultaneously to several of them. Also, changes in the input signals from the left and right neighbors can occur at many different times relative to each other and to the internal state of the cell. This task is better left to the verifier.

3.4. Verifying the Circuit

In order to verify the queue cell, we first defined flow tables for elements representing a data source and sink, which exhibit the most general behavior satisfying the above protocol. We then verify the system consisting of a single queue element with a source as its left neighbor and a sink to its right. Note that the outputs to these elements are connected to the inputs of the cell; thus, the circuit satisfies the requirement that all nodes be connected to an output. We specify a set of properties, most of which are not presented here, sufficient to ensure that the queue element would behave properly as part of a multi-cell queue.

The state graph constructed using speed-independent assumptions had 226 states. The property "if the cell is empty and IN0 is high, then IN0 stays high until 0 is stored in the cell", written

$$\mathbf{G}((\neg A \wedge \neg B \wedge C \wedge IN0) \rightarrow (IN0 \ \mathbf{U} \ (\neg A \wedge B \wedge \neg C)))$$

fails to check (as do formulas for some of the arrows in the timing diagram). Suppose that the cell is empty, and the source provides a 0 input (rasing IN0). Gate C will go low, as it should. But this lowers TCL which signals to the source that it should lower IN0. If gate B is sufficiently slow, IN0 will disappear before B has changed state, leaving all of A, B, and C low. At this point the circuit is forced to make a non-deterministic choice between raising one of them. In reality even worse situations could occur, including a metastable state between some combination of the gates or a common-mode oscillation.

It is reasonable to defend the circuit on the grounds that the source cell to the left would not be as fast as gate B, because there would be a substantial amount of logic between the TCL signal entering the source and the gate that eventually turned off IN0. This is certainly true if the source is actually another queue cell.

What happens if the source is assumed to be slower than either A or B? Adding this assumption to the circuit description results in a state graph with 72 states that satisfies all of our specifications (including those not given here). In this case, there are fewer states in the second graph partly because of the reduced non-determinism in the circuit operation.

4. Verifying High Level Descriptions of Circuits

In practice, many circuits are designed as finite state machines before they are implemented in hardware. Since this finite state machine is close to the form of state graph used by EMC, there is no need to extract it from a lower-level description of the circuit (as in the previous section). Therefore, we can verify the design before it is implemented in hardware. If a VLSI design tool that correctly implements finite state machines is used to layout the verified design, we can be sure that the resulting circuit is correct.

In order to assist with the design and verification of finite state machines, we have designed a language named SML (state machine language). In addition to being useful for verification, SML also provides a succinct notation for describing complicated finite state machines. A program written in SML is compiled into a finite state machine, which can then be verified using the model checker or implemented in hardware. At CMU, we have implemented an SML compiler that runs on a VAX 11/780. We also have access to design tools that can implement a finite state machine produced by the compiler as either a ROM, a PLA, or a PAL.

4.1. The Description Language and its Semantics

An SML program represents a synchronous circuit that implements a Moore machine. At a clock transition, the program examines its input signals and changes its internal state and output signals accordingly. Since we are dealing with digital circuits where wires are either high or low, the major data type is *boolean*. Each boolean variable may be declared to be either an *input* changed only by the external world but visible to the program, an *output* changed only by the program but visible to the external world, or an *internal* changed and seen only by the program. The hardware implementation of boolean variables may also be declared to be either active high or active low. The use of mixed logic in SML is permitted. Internal integer variables are also provided.

SML programs are similar in appearance to many imperative programming languages. SML statements include **if**, **while**, and **loop/exit**. A **parallel** is provided to allow several statements to execute concurrently in lockstep. The **break** statement can be used to terminate a **parallel** before all of the concurrent statements have finished executing. There is also a simple macro facility.

The semantics of SML programs are different from most programming languages, since we are not only interested in what a statement does, but how much time it takes to do it. In this respect, SML was influenced by the semantics of ESTEREL [2]. The complete semantics for SML will not be given here, but they will appear in a forthcoming paper [5]. A *program state* is an ordered pair, $\langle S, s\rangle$, consisting of a statement S and a function s that gives values to all of the identifiers. The semantics consist of a set of *rewrite rules* that describe how a program state can be transformed into new program state. Each rewrite rule also specifies whether it takes a clock cycle to make the transformation or not. For example, two typical rewrite rules are:

$$\langle \textbf{raise}\ (\mathrm{I});\ S,\ s\rangle \xrightarrow{1} \langle S,\ s'\rangle \quad \text{where } s' = s[\mathrm{I} \mapsto \mathit{true}]$$

$$\frac{E = \mathit{false}}{\langle \textbf{if}\ E\ \textbf{then}\ S_1\ \textbf{endif};\ S_2,\ s\rangle \xrightarrow{0} \langle S_2,\ s\rangle}$$

The first rule states that a **raise** statement followed by an arbitrary statement S can be rewritten in one clock cycle to statement S while simultaneously changing s so that $s'(\mathrm{I}) = \mathit{true}$. The second rule states that an **if** statement followed by an arbitrary statement S_2 can be rewritten in no time to statement S_2 if the condition is false.

Given any program state, we can repeatedly apply the rewrite rules to find a new state that can be reached in one clock cycle. This new state is a successor of the original state in the finite state machine. So starting from the initial program state (which consists of the entire program and a function which assigns 0 to all integers and *false* to all booleans), we can repeatedly find successor states until we have built the entire finite state machine.

4.2. Example: An ACIA Controller

The best way to illustrate the use of SML is by an example. In this example, SML will be used to design some of the controlling circuitry for an Asynchronous Communications Interface Adapter (ACIA) similar to the 6850 [15].

An ACIA converts serial data to parallel data so that it can be used by a microprocessor. The serial data is a fixed number of bits that are transmitted at a known frequency, preceded by a start bit (0) and followed by one or two stop bits (1). As the data arrives, it is shifted into a data register that can be read by the microprocessor. Once a complete byte has been received or a transmission error has been detected, the ACIA sets the appropriate bits in its status register and interrupts the microprocessor if necessary. The microprocessor can read the status register to find out what happened and then clear both registers by reading the data.

In our example, the peripheral providing the data has two connections to the ACIA, *DATA* and *DCD*. *DATA* is the input data stream of 5 bit bytes, synchronized to the ACIA controller clock by another state machine. *DCD* is the data carrier detect signal that indicates that valid data is being transmitted. If this signal goes low, the transfer has been aborted by the peripheral and the microprocessor should be interrupted so that it can find the problem.

The microprocessor has three input connections to the ACIA, *COMMAND*, *C1*, and *C0*. When *COMMAND* is high, the other two inputs indicate which function the microprocessor wants to perform. The functions are:

C1	*C0*	Function
0	0	Enable/Disable interrupt when byte has been received.
0	1	Read data register.
1	0	Read status register.
1	1	Master reset.

The ACIA controller has four outputs. When the *SHIFT* signal is high, the incoming data will be shifted into the data register. Once a complete byte has been received, the *RDRF* (read data register full) status bit should be set high. *RDRF* should be cleared if the ACIA is reset, the carrier is lost, or if the microprocessor reads the data after reading the status and finding *RDRF* high. The *IRQ* output is connected to the microprocessor to signal an interrupt request. The request is cancelled if the ACIA is reset or if the microprocessor reads the status and data registers. The *DCDO* (data carrier detect outstanding) status bit is set high whenever there has been a loss of carrier that hasn't been handled by the microprocessor. It is set high as soon as the *DCD* input goes low. It remains high as long as *DCD* remains low. If the microprocessor either reads the status and data registers or resets the ACIA while *DCD* is low, *DCDO* will go low as soon as the carrier is restored. Otherwise, *DCDO* will remain high until the status and data registers are read or the ACIA is reset.

4.3. An Implementation of the ACIA Controller in SML

```
program acia;

input C0, C1, COMMAND, DCD, DATA;
output SHIFT, RDRF, IRQ, DCDO;
internal NEWIRQ, DCDST, DCDRD, RIE;

#define RESET          (COMMAND & C1 & C0)
#define READSTAT       (COMMAND & C1 & !C0)
#define READDATA       (COMMAND & !C1 & C0)

procedure wait(exp)
    while !(exp) do loop skip endloop
endproc

    parallel
        loop
            parallel
                loop
                    if !DCD | RESET then break endif
                endloop
            ||
                loop
                    wait(!DATA);
                    raise(SHIFT);
                    delay 4;
                    lower (SHIFT);
                    parallel
                        if RIE then
                            parallel raise(IRQ) || raise(NEWIRQ) endparallel
                        endif
                    ||
                        raise (RDRF)
                    endparallel
                endloop
            endparallel;
            parallel lower(SHIFT) || lower(RDRF) endparallel
        endloop
    ||
        loop
            wait(!DCD);
            parallel
                raise(DCDO)
            ||
                if !RESET then
                    parallel raise(IRQ) || raise(NEWIRQ) || raise(DCDST) || raise(DCDRD) endparallel
                endif
            ||
                wait(DCD)
            endparallel;
            DCDO := DCDRD;
        endloop
    ||
        loop
            wait(COMMAND);
            switch
                case READSTAT:
                    parallel lower(NEWIRQ) || lower(DCDST) endparallel;
                    break;
                case READDATA:
                    parallel
                        lower(RDRF) || IRQ := NEWIRQ || DCDRD := DCDST
                    endparallel;
                    break;
                case RESET:
                    parallel lower(IRQ) || lower(NEWIRQ) || lower(DCDST) || lower(DCDRD) endparallel;
                    break;
                default:        -- change read interrupt enable
                    invert(RIE);
                    break;
            endswitch
        endloop
    endparallel
endprog
```

Figure 4-1: A First Attempt at Writing the ACIA Controller in SML

Figure 4-1 shows a program that attempts to implement the ACIA controller. The numbers at the beginning of each line were added for easy reference and are not part of the language.

A few comments are necessary to explain the operation of this program.

Lines 3-5: In addition to declaring the input and output signals, four internal booleans are also declared. *NEWIRQ* is true when an interrupt has been generated but the microprocessor has not discovered why by reading the status register. *DCDST* is true when the carrier has been lost but the microprocessor hasn't read the status register to discover the problem. *DCDRD* is true when the carrier has been lost and the microprocessor has not handled this by reading the status register and then the data register. *RIE* (receive interrupt enable) is true if the microprocessor should be interrupted when a complete byte has been received.

Lines 7-9: The SML compiler includes a preprocessor similar to the C preprocessor. These lines define three macros that are used to decode the command that the microprocessor is issuing.

Lines 11-13: **Wait** is defined to be a macro that does nothing until its argument becomes true.

Lines 22-34: This loop handles the incoming data. The procedure is to wait for the start bit (line 23) and then start shifting (line 24). After 4 more bits have been received, stop shifting (lines 25 and 26). The data register is now full, and an interrupt should be sent if interrupts are enabled (lines 27-33). The loop is restarted, and the ACIA waits for the next start bit.

Lines 18-20: At any time, it is possible for the carrier to go away or the microprocessor to execute a reset. Should this happen, the current transfer should be aborted. This loop waits for either event to occur, and then breaks out of the **parallel**, thus terminating the execution of the loop at lines 22-34.

Line 36: If the carrier is lost or the ACIA should be reset, the break on line 19 causes execution to resume here. This statement stops shifting and empties the data register. Then the normal incoming data loop is restarted.

Lines 39-51: This loop handles the loss of carrier. To begin with, the controller waits for the carrier to go away (line 40). Once the carrier has been lost, the *DCD* status bit is set (line 42). If the ACIA is not being reset, the ACIA interrupts the microprocessor and indicates that a new loss of carrier has been detected (lines 44-46). Then the ACIA waits for the carrier to be restored (line 48). Once the carrier has been restored, the *DCD* status bit is cleared (line 50) if the microprocessor has done something about the loss of carrier (and *DCDRD* has been reset as a result).

Lines 53-71: This loop handles requests from the microprocessor. Nothing is done until a valid command is received (line 54). If the command is a read status, the microprocessor will know about any interrupt or loss of carrier, so *NEWIRQ* and *DCDST* are cleared (lines 56-58). If the command is a read data, the data register has been emptied, so RDRF should be cleared (lines 59-63). In addition, if a prior read status has found out about an interrupt (*NEWIRQ* is low), the interrupt should be cancelled. If a prior read status has found out about a loss of carrier (*DCDST* is low), then we should indicate that the loss has been handled by lowering *DCDRD*. If the command is a reset, the ACIA cancels any interrupt and ignores a simultaneous loss of carrier (lines 64-66). If the command is enable/disable interrupts, the current value of *RIE* is changed (lines 67-69). After processing the command, the loop is restarted and the ACIA waits for the next command.

This program was compiled into a minimal 450 state machine in approximately 25 seconds of CPU time on a VAX.

4.4. Verifying the ACIA Controller Using the Model Checker

The correct behavior of a controller such as the ACIA is typically described by a set of timing diagrams, such as the one in figure 4-2. In this section, we describe how this timing diagram can be expressed in CTL.

In order to simplify the CTL description of this timing diagram, we introduce a set of macro definitions. Since *DCD*, *DCDO* and *IRQ* are usually high and *COMMAND* is usually low, we define **normal** to be $DCD \wedge DCDO \wedge IRQ \wedge \neg COMMAND$. In addition, **readstat** is defined to be $COMMAND \wedge C1 \wedge \neg C0$ and **readdata** is defined to be $COMMAND \wedge \neg C1 \wedge C0$. Using these simple definitions, we introduce a macro **nextcommand**(*type,next*) that is true if along all paths where *DCD* remains high and the next microprocessor request is *type*, *DCDO* and *IRQ* remain high and *next* becomes true after the command is received. The definition of **nextcommand** is:

$$\textbf{nextcommand}(type,next) := [\textbf{normal u } \neg\textbf{normal} \wedge (DCD \wedge type \rightarrow DCDO \wedge IRQ \wedge \textbf{X}\ next)]$$

In English, this formula states that along all paths, *DCD*, *DCDO* and *IRQ* are high and *COMMAND* is low until something changes. At this point, if *DCD* is still high and the microprocessor has issued a *type* command, then *DCDO* and *IRQ* should remain high and *next* should be true in all next states.

Now that we have defined **nextcommand**, it is a simple matter to describe this timing diagram in CTL by the formula:

$$\textbf{G}\,(\neg DCD \wedge \neg COMMAND \wedge \neg DCDO \wedge \neg RIE \rightarrow$$

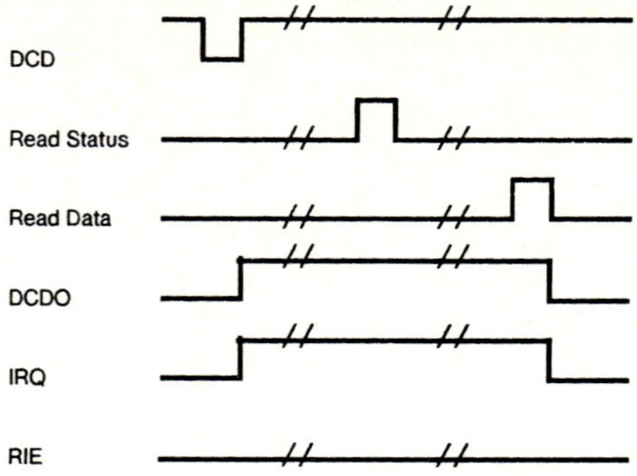

Figure 4-2: A Sample Timing Diagram for the ACIA

```
% emc -c acia1.asm
                CTL MODEL CHECKER (version B1.0)

Reading acia...
Fairness constraint: .

time:  (1066 244)

|= G (~DCD -> X DCDO).
The equation is TRUE.


time:  (338 89)

|= G (~DCD & ~COMMAND & ~DCDO & ~RIE ->
      X nextcommand(readstat, nextcommand(readdata, ~DCDO & ~IRQ))).
The equation is FALSE along the path:
State 0:
State 1: C1 COMMAND DCD NEWIRQ DCDST DCDRD IRQ DCDO
State 377: C0 COMMAND DCD DCDRD SHIFT IRQ DCDO
State 250: COMMAND DCD SHIFT DCDO

time:  (2392 536)

|= .
End of session.
%
```

Figure 4-3: Verifying the First ACIA Controller Program

```
% emc -c acia2.asm
                CTL MODEL CHECKER (version B1.0)

Reading acia...
Fairness constraint: .

time:  (813 252)

|= G (~DCD -> X DCDO).
The equation is TRUE.


time:  (275 66)

|= G (~DCD & ~COMMAND & ~DCDO & ~RIE ->
      X nextcommand(readstat, nextcommand(readdata, ~DCDO & ~IRQ))).
The equation is TRUE.


time:  (1670 581)

|= .
End of session.
%
```

Figure 4-4: Verifying the Corrected ACIA Controller Program

X nextcommand(*readstat*, **nextcommand(***readdata*, $\neg DCDO \wedge \neg IRQ$)))

At the begining of figure 4-2, *DCD*, *COMMAND*, *DCDO*, and *RIE* are all low. Whenever this situation occurs, *DCDO* and *IRQ* will go high and remain high until a read status command is received (this is the meaning of **nextcommand**(readstat,...)). After the read status, *DCDO* and *IRQ* will remain high until a read data command is received (according to the nested **nextcommand**). After the read data, *DCDO* and *IRQ* will both go low.

Figure 4-3 shows a transcript of the model checker running on the program in figure 4-1. The numbers in parentheses are the total user cpu time and "system time", in 1/60ths of a second.

As the transcript shows, the program does raise *DCDO* whenever *DCD* goes high, but it doesn't lower *DCDO* if *DCD* goes low before a read data occurs. Therefore, we change lines 59-63 of the original program to be:

```
59          case READDATA:
60              parallel
61                  lower(RDRF) || IRQ := NEWIRQ || DCDRD := DCDST || DCDO := DCDST | !DCD
62              endparallel;
63              break;
```

With this change, *DCDO* will be lowered if *DCD* is high and *DCDST* is low (indicating that a read status has occurred). The new program compiles into 352 states in approximately 20 seconds of CPU time. The correctness of this program is shown by the transcript in figure 4-4.

5. Directions For Future Research

Some circuits seem easier to specify by using a timing diagram like figure 3-3 than by a CTL formula. Of course, CTL is more general since there is no analogue of negation, disjunction, or conjunction for timing diagrams. It may be possible to either systematically translate timing diagrams into CTL formulas or check them directly using an algorithm similar to the one used by EMC. If so, this would simplify the task of specifying a complicated circuit and also allow the designer to be more confident that specifications actually mean what he thinks they mean.

In the case of asynchronous circuits more research is needed on realistic timing models. Clearly the arbitrary delay model used in [11] is too conservative. The use of delay constraints as described in section 3 is a step in the right direction. But it still does not handle timing assumptions like the 3/2 rule, which requires that the delay through any three gates be greater than the delay through any two. A more general scheme for delay constraints that will handle the 3/2 rule and other similar timing assumptions is under development.

More research is also needed on how to extend the approach described in this paper to large, hierarchically constructed circuits. A first step in this direction is discussed in [8]. If one uses a subset of CTL, lower level circuits can be simplified by "hiding" some of their internal nodes (more precisely, making it illegal to use them in CTL formulas) and merging groups of states that become indistinguishable into single states.

Finally, we believe in practice it may be possible to verify circuits that consist of arrays of identical cells by using the model checker to verify a generic instance of the cell design and then applying some form of induction. Because CTL is propositional, however, the induction rule will probably have to be formulated outside of the logic.

References

[1] M. Browne, E. Clarke, D. Dill, B. Mishra, "Automatic Verification of Sequential Circuits", *CHDL85*, Tokyo, August 1985.

[2] G. Berry and L. Cosserat, "The ESTEREL Synchronous Programming Language and its Mathematical Semantics", Tech. report, Ecole Nationale Superieune des Mines de Paris, 1984.

[3] M. Ben-Ari, Z. Manna, A. Pneuli, "The Logic of Nextime", *8th ACM Symposium on Principles of Programming*, Jan. 1981.

[4] G. V. Bochmann, "Hardware Specification with Temporal Logic: An Example", *IEEE Transactions on Computers*, Vol. C-31, No. 3, March 1982.

[5] M.C.Browne, E.M.Clarke, "SML-A Finite State Language".

[6] E.M. Clarke, E.A. Emerson, "Synthesis of Synchronization Skeletons for Branching Time Temporal Logic", in *Proc. of the Workshop on Logic of Programs, Yorktown Heights, NY*, 1981.

[7] E.M. Clarke, E.A. Emerson, A.P. Sistla, "Automatic Verification of Finite-State Concurrent Systems using Temporal Logic Specifications: A Practical Approach", *Tenth ACM Symposium on Principles of Programming Languages*, Austin,Texas, 1983.

[8] E.M. Clarke and B. Mishra, "Automatic Verification of Asynchronous Circuits", in *Proc. of Logics of Programs, Pittsburgh. Pa.*, E. Clarke and D. Kozen, eds., 1983.

[9] George Cogar, "Asynchronous Self Controlled Shift Register", U.S. Patent Office 3,166,715. January, 1965.

[10] David L. Dill, "Verification of Asynchronous Circuits with Timing Dependencies".

[11] David L. Dill and Edmund M. Clarke, "Automatic Verification of Asynchronous Circuits using Temporal

Logic", *1985 Chapel Hill Conference on VLSI*, Computer Science Press, May 1985.

[12] E.A. Emerson, E.M. Clarke, "Characterizing Properties of Parallel Programs as Fixpoints", in *Proc. of the Seventh International Colloquium on Automata, Languages and Programming*, 1981.

[13] Y. Malachi and S. S. Owicki, "Temporal Specifications of Self-Timed Systems", *VLSI Systems and Computations*, 1981.

[14] A. Pneuli, "The Temporal Semantics of Concurrent Programs", *18th Symposium on Foundations of Computer Science*, 1977.

[15] Veronis, Andrew, *Microprocessors: Design and Applications*, , 1978.

Formal Aspects of VLSI Design
G.J. Milne and P.A. Subrahmanyam (editors)

Can a Simulator Verify a Circuit?

Randal E. Bryant

Dept. of Computer Science
Carnegie-Mellon University

A logic simulator can prove the correctness of a digital circuit if only circuits implementing the system specification can produce a particular response to a sequence of simulation commands. This paper explores two methods for verifying circuits by a three-valued logic simulator where the third state X indicates an indeterminate value. The first, called "black-box" simulation, involves simply observing the output produced by the simulator in response to a sequence of inputs with no consideration of the internal circuit structure. This style of simulation can verify only a limited class of systems. The second method, called "transition" simulation, requires the user to specify the relation between states in the circuit and the specification. The simulator is then used to prove that each state transition in the specification is implemented correctly. Arbitrary systems may be verified by this method.

Introduction

Digital designers rely primarily on logic simulators to test the correctness of their logic circuits. Typically, however, only a limited set of test cases are simulated, and the circuit is presumed correct if the simulator yields the expected results for all of these cases. This style of simulation is analogous to the unstructured way most programmers test programs. To paraphrase a saying of Dijkstra, unstructured use of a simulator can detect the presence of errors but can never prove their total absence. Stories abound of circuit design errors that remain undetected despite many hours of simulation time. Given the great effort required to design and manufacture a system of VLSI complexity, the cost of undetected design errors can be quite high. For this reason more reliable methods of circuit verification are desired.

Consider the possibility of rigorously verifying a circuit by logic simulation. If it could be shown that only circuits meeting the specification will produce satisfactory results for a given set of test cases, then a simulator could serve as a fully automated verification system. Since logic simulators have been developed for a wide range of circuit models and digital design styles, such an approach would not require extensive development of a new class of programs as would other verification methods. Logic simulators seem like a good starting point for developing a circuit verification program. By investigating their expressive power, we may better understand what additional capabilities are required. Thus, we are motivated to ask: "Under what conditions does a simulation constitute a proof of the circuit's correctness?"

Before answering this question, we must formulate more precise definitions of the system specification, the circuit, and the simulator. We do this in a somewhat abstract way such that the results apply to a wide variety of circuit technologies and simulators. The desired behavior is specified in terms of a finite state automaton. A circuit is also described as a finite state automaton, but with a particular binary encoding of the states. The circuit is said to implement the specification if the two automata are input–output equivalent. The simulator models the behavior of the circuit automaton, computing new state and output values in response to inputs supplied by the user. The simulator extends the normal logic values 0 and 1 with a third value, denoted X, representing an unknown or indeterminate condition. The simulator produces responses that are monotonic with respect to the ordering $X < 1$ and $X < 0$. That is, when some of the input or state variables equal X, the circuit is simulated as if it could have any input and state obtained by replacing each X with 0 or 1. The simulator may produce output or new state 0 or 1 only if the circuit would produce this response for all states and inputs represented by the state and input variable values. During simulation, the user can command the simulator to *Erase* the internal state, i.e. set all state variables to X. This gives the user a limited and indirect capability to control and observe the internal state of the simulation. The simulator is assumed to correctly model the behavior of the circuit. In practice this assumption does not always hold, such as when a subtle electrical problem, beyond the modeling capabilities of the simulator, would cause the circuit to malfunction. Like any abstract verification method, ours is valid only as long as the actual system satisfies these abstractions.

The task of verifying a circuit with a simulator bears many similarities to the "machine identification" problem of sequential systems theory[8]. Given a sequential system starting in an arbitrary state, there is no way to "identify" it, i.e. establish its equivalence to a specified automaton, by simply observing its response to a sequence of stimuli. On the other hand, given additional information about the system such as an upper bound on the number of system states, the problem becomes solvable. In our formulation, we assume no bound on the number of circuit states but instead assume additional capabilities of the simulator, namely three-valued modeling. In many circuit technologies, such as those that allow values to be stored as charge on the wires[7], there is no straightforward method to identify all storage points in a circuit. Without tight bounds on the number of circuit states, the methods developed in the context of machine identification do not appear viable for circuit verification. On the other hand, exploiting the modeling capabilities found in most logic simulators allows us to resolve the identification problem in a very natural way.

Any method of verification can err in two different ways. When the verifier rejects a correct circuit, its response is *false negative*, whereas a *false positive* response occurs when it accepts an incorrect circuit. We will concern ourselves primarily with avoiding false positive responses. Such a response has more potential for danger —it may cause a defective design to be implemented or put into service. Furthermore, deciding whether a simulator has produced a false negative response requires more detailed information about the circuit electronics and the simulation algorithm than can be presented in a general way. However, for the simulation sequences presented in this paper, a false negative response must have a particular form, namely the simulator will produce X on some output when 0 or 1 was expected. We also impose a nontriviality condition on the verifier so that it cannot simply reject all circuits as a means of never giving a false positive response.

This paper investigates two approaches to circuit verification by simulation. In the first, the user must view the simulated circuit as a "black box", defined only by its input–output behavior with no knowledge of the circuit structure or even the number of state variables. No direct observation or modification of circuit state is permitted. This approach has the advantage that the simulation sequence need not depend on the circuit design. It can be shown that a circuit can be verified by black-box simulation if and only if the specified system is *definite*. That is, for some constant k, the output of the system at any time depends only on the last k inputs. For example, combinational circuits ($k = 1$) and shift registers ($k =$ the shift register length) are definite. On the other hand, ordinary registers and most other sequential systems are not definite, and hence a black-box simulator does not have sufficient expressive power to verify them. In the second simulation method, the user must specify the relation between the specification and circuit states. The simulator is then used to verify that each circuit state produces the desired output and that each state transition is implemented correctly. This approach can verify arbitrary systems, but the simulation sequence now becomes circuit

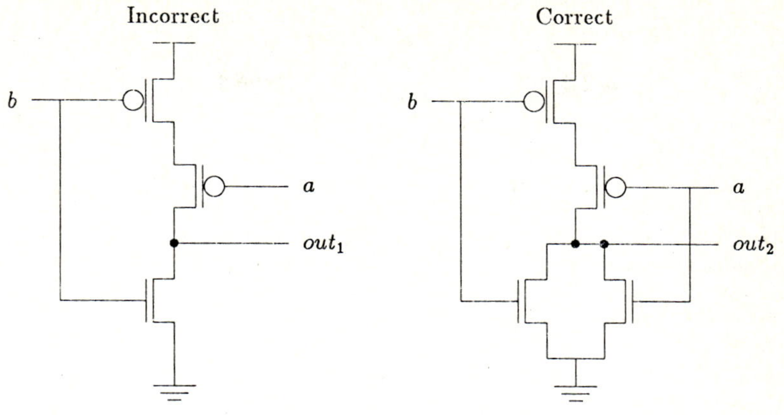

Figure 1: Implementations of a Nor Gate in CMOS

dependent and requires greater development effort on the part of the user.

Verification Example

To further motivate the difficulty of verifying a circuit by simulation, and to preview some results of this paper, consider the seemingly simple task of proving that a circuit implements a "Nor" logic gate. Figure 1 shows two proposed implementations in CMOS technology [4]. If we were to simulate these circuits using a simulator that can model a MOS circuit at the transistor level[2], the following responses would be produced when the input patterns are applied in the sequence shown:

a	b	out_1	out_2
0	0	1	1
0	1	0	0
1	1	0	0
1	0	0	0

The two circuits appear identical for all possible input combinations, as they would be in actual implementations of the circuits. However, only the second circuit is a valid Nor gate. The first is a two state sequential circuit, because when $a = 1$ and $b = 0$, the output node is electrically isolated from all others and remains charged at its previous value. Due to the order in which the input combinations were applied, it just happened that the previous value of out_1 equaled the value a Nor gate should produce for this input combination. On the other hand, had this input combination been simulated immediately after the combination $a = b = 0$, the output would have been 1.

This example illustrates a common problem in testing a circuit by simulation even when rigorous verification is not sought. A design error introduces an unintended sequential dependency in the circuit, but this error remains undetected because of the particular order in which the test sequences are simulated. Clearly, such a condition is not acceptable for formal verification.

Suppose, on the other hand, that before each input combination is simulated, all state variables are set to X. Such a simulation would produce the following results:

a	b	out_1	out_2
0	0	1	1
0	1	0	0
1	1	0	0
1	0	X	0

The presence of an X on out_1 indicates that the output of this circuit may not be uniquely defined when $a = 1$ and $b = 0$ (or it could be a false negative response by a valid circuit). On the other hand, each input combination produces unique response for the second circuit, and since the responses match those of a Nor gate, one can conclude the circuit is correct. This conclusion can be drawn without any further information about the structure of the circuit or the number of state variables.

A combinational system such as a Nor gate is 1-definite: its output at any time depends only on the most recent input. The method shown above generalizes to any specification where the output depends only on the most recent k inputs for some constant k. That is, suppose for every possible input sequence of length k, setting all state variables to X and then simulating the sequence yields an output equal to the desired value. One can then safely conclude the circuit implements the specification.

On the other hand, many sequential systems are not definite. For example, a simple 1-bit register has an output that can depend on an input that occurred arbitrarily long in the past if no new data are written into it. For such a system, given any value k, there will always be an input sequence of length k that does not cause the system to produce a unique output. When simulating this sequence with all state variables initialized to X, even a correctly designed circuit will give an X on the output. We will show that for an indefinite system, there is no complete set of input sequences for validating a circuit by simply observing the output values during simulation.

Problem Formulation

This section develops a mathematical abstraction of logic circuits, simulators, and the verification problem.

Notation

We adopt a notation that represents the system input, output, and state values as vectors and their behavior over time as sequences of vectors.

- p — The number of system inputs
- B — $\{0, 1\}$
- B^n — $\{\langle x_1, \ldots, x_n\rangle | x_i \in B\}$
- Φ_l — All length l sequences with elements in B^p
- Φ — $\bigcup_{0 \leq l < \infty} \Phi_l$
- V — $\{0, 1, X\}$, partially ordered $X < 1$ and $X < 0$.
- V^n — $\{\langle x_1, \ldots, x_n\rangle | x_i \in V\}$, partially ordered $\vec{x} \leq \vec{y}$ if $x_i \leq y_i$ for all $1 \leq i \leq n$.
- Ψ_l — All length l sequences with elements in V^p.
- Ψ — $\bigcup_{0 \leq l < \infty} \Psi_l$, partially ordered $[\vec{a}_1, \ldots, \vec{a}_s] \leq [\vec{b}_1, \ldots, \vec{b}_t]$ if $s \leq t$ and $\vec{a}_{s-i} \leq \vec{b}_{t-i}$ for all $0 \leq i < s$.
- ϵ — The empty sequence.
- $\alpha \cdot \beta$ — The concatenation of sequences α and β.

The partial ordering $X < 0$ and $X < 1$ orders values by their "information content." That is, X indicates an absence of information while 0 and 1 represent specific, fully-defined values. When speaking of domains ordered by information content we shall term values a and b "consistent" if either $a \leq b$ or $b \leq a$, and "inconsistent" otherwise.

The information ordering is extended to vectors and sequences of vectors by adopting the convention that one value is less than another if the elements of the first are consistent with those of the second, but the first contains less information than the second. More precisely, for vectors in V^n, one vector is less than or equal to another if each element of the first is less than or equal to each element of the second. For sequences $\alpha, \beta \in \Psi$, α is less than or equal to β if the elements of α are less than or equal to the corresponding final elements of β. That is, the history given by α is consistent with the most recent history given by β, but may have less information content. For $\alpha, \beta \in \Phi$, i.e. sequences of Boolean vectors, $\alpha \leq \beta$ when α is a suffix of β.

Definition 1: For partially ordered sets D_1, D_2 a *monotonic* function $g: D_1 \rightarrow D_2$ satisfies

$$a \leq b \Longrightarrow g(a) \leq g(b)$$

for all $a, b \in D_1$. Similarly, a monotonic function with multiple arguments satisfies this property for each argument.

Monotonicity describes in mathematical terms a desirable property of systems that process data ordered by information content. That is, suppose a system's response to a stimulus is represented by a monotonic function. If one stimulus is consistent with another but possibly contains less information, the respective responses they cause will preserve this ordering.

Definition 2: For partially ordered sets D_1 and D_2 with subsets D_1' and D_2', respectively, a *monotonic extension* of function $f: D_1' \rightarrow D_2'$ is a function $g: D_1 \rightarrow D_2$ such that g is monotonic and $f(a) = g(a)$ for all $a \in D_1'$. Similarly, a monotonic extension of a multiple argument function must satisfy these properties for each argument.

As an example, the "or" function can be extended monotonically from the Boolean to the ternary domain in several different ways, including the following:

$$a \vee b = \begin{cases} 1, & a = 1 \text{ or } b = 1 \\ 0, & a = 0 \text{ and } b = 0 \\ X, & \text{else.} \end{cases} \tag{1}$$

Other extensions yield X when $a = 1$ and $b = X$ or *vice-versa*.

System Specification

The system to be implemented has p inputs and m outputs, each of which may equal 0 or 1. Hence the system may be described as a finite automaton with input alphabet B^p and output alphabet B^m.

Definition 3: A system specification $\mathcal{M}$ is a triple $\langle Q, Next, Out \rangle$ with

Q: a finite set of states,

$Next$: the next state function $Next: Q \times B^p \rightarrow Q$, and

Out: the output function $Out: Q \rightarrow B^m$.

The function $Next$ is extended to input sequences to yield the state after all inputs have been applied, i.e. to a function $Next: Q \times \Phi \rightarrow Q$ defined recursively as

$$\begin{aligned} Next(q, \epsilon) &= q \\ Next(q, \alpha \cdot \vec{x}) &= Next(Next(q, \alpha), \vec{x}). \end{aligned}$$

The function $FinalOut: Q \times \Phi \rightarrow B^m$ is defined to yield the final output after a sequence of inputs has been applied, i.e.

$$FinalOut(q, \alpha) = Out(Next(q, \alpha)).$$

Circuit Model

A circuit is also a finite automaton but with states encoded by s Boolean variables.

Definition 4: A circuit $\mathcal{C}$ is a triple $\langle s, next, out\rangle$ with

s: the number of state variables,

$next$: the next state function $next: B^s \times B^p \to B^s$, and

out: the output function $out: B^s \to B^m$.

Definition 5: Circuit $\mathcal{C}$ *implements* specification $\mathcal{M}$ when there exists a relation $\mathcal{E} \subseteq Q \times B^s$ (for "encodes") satisfying:

1. For every $q \in Q$ there exists a $\vec{z} \in B^s$ for which $q\ \mathcal{E}\ \vec{z}$.
2. For any $q \in Q$ and $\vec{z} \in B^s$
$$q\ \mathcal{E}\ \vec{z} \implies Out(q) = out(\vec{z}),$$
3. For any $q \in Q$, $\vec{z} \in B^s$, and $\vec{x} \in B^p$,
$$q\ \mathcal{E}\ \vec{z} \implies Next(q, \vec{x})\ \mathcal{E}\ next(\vec{z}, \vec{x}).$$

Remark: By this definition, the circuit automaton must "cover" the input–output behavior of the specification. That is, for any initial state of the specification, there must be an initial circuit state, such that the two automata would yield identical outputs for any subsequent input sequence. However, there may be circuit states that do not correspond to any specification states, such as those involving invalid combinations of state variables. Furthermore, the automata need not be reduced —several circuit states may correspond to a single specification state and *vice-versa*.

Simulator

For a circuit $\mathcal{C} = \langle s, next, out\rangle$, the simulator maintains state variables $\vec{z} \in V^s$, and computes the behavior according to functions $next: V^s \times V^p \to V^s$ and $out: V^s \to V^m$, which are arbitrary monotonic extensions of the corresponding circuit functions. The simulator implements five commands, although for black-box simulation only the first three are allowed.

Erase: causes the simulator to set z_i to X for all $i \in \{1, \ldots s\}$.

Cycle($\vec{x}$): causes the simulator to set $\vec{z}$ to $next(\vec{z}, \vec{x})$.

Output: causes the simulator to print $out(\vec{z})$.

Set(i, b): causes the simulator to set z_i to b for $b \in V$.

Observe(i): causes the simulator to print z_i.

Simulation Experiment

Definition 6: A *simulation experiment* consists of a sequence of simulation commands beginning with Erase, as well as a procedure by which the user decides whether the outcome is acceptable. The decision procedure can depend only on the values produced by the Output and Observe commands.

Definition 7: An experiment is *effective* for specification $\mathcal{M}$ when the simulation of a circuit $\mathcal{C}$ can be judged acceptable only if $\mathcal{C}$ implements $\mathcal{M}$.

Remark: This condition requires that the verification method cannot produce any false positive responses.

Definition 8: An experiment is *nontrivial* when the simulator produces an acceptable outcome for some circuit $\langle s, next, out\rangle$ and monotonic extensions of $next$ and out.

Remark: This condition is imposed to eliminate the otherwise effective test of rejecting all circuits.

Black-Box Simulation

With black-box simulation, the user is limited to the simulation commands Erase, Cycle, and Output. No direct observation or modification of the simulator state $\vec{z}$ is permitted. This section identifies the class of systems that can be verified by black-box simulation.

Define the function $SimState: \Psi \to V^s$ as the state of the simulator after giving an Erase command followed by a series of Cycle commands. More precisely

$$\begin{aligned} SimState(\epsilon) &= X^s \\ SimState(\alpha \cdot \vec{x}) &= next(SimState(\alpha), \vec{x}) \end{aligned}$$

where X^s denotes a vector with each element equal to X.

Similarly, define the function $SimOut: \Psi \to V^m$ as the result that would be printed by an Output command following the simulation of some input sequence, i.e.

$$SimOut(\alpha) = out(SimState(\alpha)).$$

A black-box simulation experiment can be viewed as a decision procedure that either accepts or rejects a circuit based on the values of $SimOut(\alpha)$ for any finite number of sequences $\alpha \in \Psi$.

Definite Systems

Definition 9: For any $k \geq 0$, specification $\mathcal{M}$ is k-definite if $FinalOut(q_1, \alpha) = FinalOut(q_2, \alpha)$ for any $\alpha \in \Phi_k$ and any $q_1, q_2 \in Q$,

Kohavi[6] defines a k-definite system as one for which any input sequence of length k places the system in a unique state, whereas we only require the sequence to cause a unique output. If the specification automaton is in reduced form, the two definitions are equivalent.

Definition 10: A specification is *definite* if it is k-definite for any k. Otherwise it is *indefinite.*

Monotonicity Properties

In this section we will state several properties of the logic simulator that follow from the monotonicity of the simulator functions *next* and *out*. As shall be seen, monotonicity provides the primary mechanism by which one can guarantee properties of a circuit knowing only its responses during simulation.

Proposition 1: The functions *SimState* and *SimOut* are monotonic.

This proposition shows how the Erase command and the three-valued modeling enhances the power of the simulator. If an Erase command followed by a sequence of Cycle commands causes the simulator to produce a 0 or 1 on some output or state variable, then this sequence of inputs must also cause the circuit to produce the same output or state regardless of the initial state.

Proposition 2: If circuit C implements specification $\mathcal{M}$ then for all $\alpha \in \Psi$ and all $\beta \in \Phi$ such that $\alpha \leq \beta$:

$$SimOut(\alpha) \leq FinalOut(q, \beta)$$

for all $q \in Q$.

The proof that a k-definite specification can be verified by simulating it for all input sequences of length k relies on the following proposition.

Proposition 3: For any $k \geq 0$, if $SimOut(\alpha) = FinalOut(q, \alpha)$ for all $\alpha \in \Phi_k$ and all $q \in Q$, then the simulated circuit C implements specification $\mathcal{M}$.

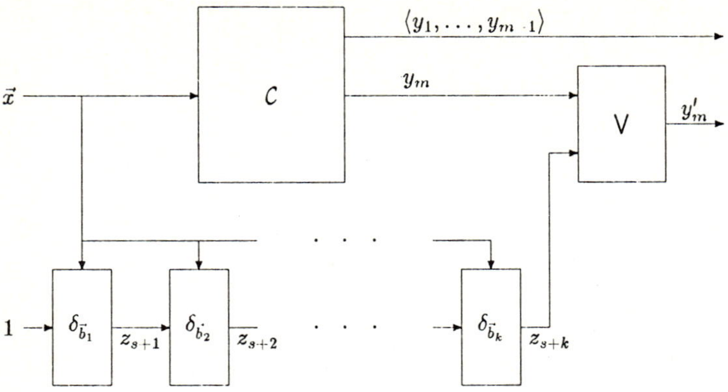

Figure 2: Circuit with Booby Trap

Expressive Power

Theorem 1: There exists an effective, nontrivial, black-box simulation experiment for specification $\mathcal{M}$ if and only if $\mathcal{M}$ is definite.

Outline of Proof: First, suppose $\mathcal{M}$ is k-definite for some value k. Consider the simulation experiment consisting of executing the sequence of commands required to compute $SimOut(\alpha)$ for each $\alpha \in \Phi_k$, and accepting the circuit if $SimOut(\alpha) = FinalOut(q, \alpha)$ for all α and any choice of $q \in Q$ (in a k-definite specification, the choice of initial state makes no difference.) Proposition 3 shows that this experiment is effective.

Furthermore, it is a straightforward exercise to implement a definite system using shift registers to store the most recent k input patterns and combinational logic to compute the output function [6]. By appropriate monotonic extensions of the circuit functions, we obtain a circuit for which $SimOut(\alpha) = FinalOut(q, \alpha)$ for any $q \in Q$. Therefore, the simulation experiment is nontrivial.

Next, suppose $\mathcal{M}$ is not k-definite for any value of k. We use an adversary argument to show that no nontrivial, black-box experiment for this specification can also be effective. Assume that there is some nontrivial simulation experiment in which fewer than k Cycle commands occur without an intervening Erase command, for some value k. Since the experiment is nontrivial, there must be a circuit $\mathcal{C} = \langle s, next, out \rangle$ and monotonic extensions of $next$ and out that produce an acceptable result for the experiment. We will construct a circuit $\mathcal{C}' = \langle s', next', out' \rangle$ that does not implement $\mathcal{M}$. However, we can define monotonic extensions of $next'$ and out' such that by defining $SimOut'$ in a manner analogous to the definition of $SimOut$, we have $SimOut'(\alpha) = SimOut(\alpha)$ for any sequence $\alpha \in \Psi_l$, for which $l < k$. The simulation experiment cannot possibly distinguish $\mathcal{C}$ from $\mathcal{C}'$ and hence is not effective. This argument holds for any value of k, showing that there is no finite upper bound on the length of a simulation experiment that can distinguish a correct circuit from a defective one when the specified system is indefinite.

Circuit $\mathcal{C}'$ is constructed as illustrated in Figure 2 by taking circuit $\mathcal{C}$ and adding extra logic to implement a "booby trap", i.e. logic that will not affect the output value until some input sequence of length k occurs. Designing such a booby trap is no easy task, because any state variables used by the trap will be set to X whenever the user gives an Erase command. For an improper design this could cause $\mathcal{C}'$ to produce an X on its output under conditions when $\mathcal{C}$ would not. Similarly, the user might attempt to expose any traps by presenting inputs with some elements equal to X.

For the design of $\mathcal{C}'$, let $\beta = [\vec{b}_1, \ldots, \vec{b}_k] \in \Phi_k$ be some sequence such that $\mathcal{M}$ has states

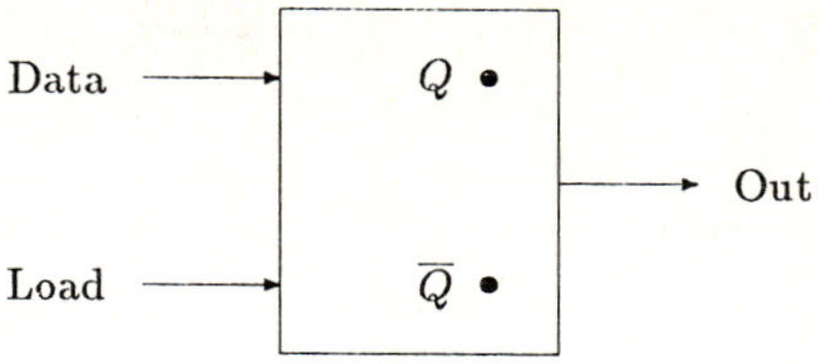

Figure 3: Register Circuit

$q_1, q_2 \in Q$ for which $FinalOut(q_1, \beta) \neq FinalOut(q_2, \beta)$. Such a sequence must exist or else $\mathcal{M}$ would be k-definite. Assume for simplicity, that output m differs for these two cases, relabeling the outputs if required. The trap consists of a shift register, where each shift element sets its output to its input value when the circuit input matches the corresponding element of β and clears the output to 0 otherwise. Consequently, input sequence β will cause a 1 to propagate through the shift register, forcing output m to 1 when it reaches the end. Any input sequence $\alpha \not\leq \beta$ of length less than or equal to k will cause the shift register to produce 0, leaving the circuit output unchanged. The behavior of the circuit for input sequences $\alpha < \beta$ will depend on the initial state, and hence under such conditions $SimOut'_m(\alpha) = X$. However, it can be shown using Proposition 2 that under these conditions $SimOut_m(\alpha) = X$ as well.

□

State Transition Simulation

To verify circuits implementing a more general class than definite systems, the results of the previous section imply that a capability beyond black-box simulation is required. More information about the internal structure of the circuit must be introduced, such as how states in the specification are encoded by state variables in the circuit states. We would prefer, however, to introduce as little information as possible about the circuit structure into the verification. Toward this goal we will develop a notation similar to the Floyd-Hoare assertion method of program verification, along with an associated simulation methodology for testing assertions.

Definition 11: A *circuit assertion* is a set of four predicates:

$P(\vec{z})$: a precondition on the state variables, $P\colon B^s \rightarrow \{true, false\}$.

$I(\vec{x})$: a condition on the input variables, $I\colon B^p \rightarrow \{true, false\}$.

$N(\vec{z})$: a postcondition on the state variables, $N\colon B^s \rightarrow \{true, false\}$.

$O(\vec{y})$: a postcondition on the output variables, $O\colon B^m \rightarrow \{true, false\}$.

Remark: A circuit assertion is denoted by an equation of the form the form $P\{\ I\ \}N \wedge O$.

Definition 12: A circuit satisfies assertion $P\{\ I\ \}N \wedge O$ if $N[next(\vec{x}, \vec{z})]$ and $O[out(next(\vec{x}, \vec{z}))]$ for all $\vec{z} \in B^s$ such that $P(\vec{z})$ and all $\vec{x} \in B^p$ such that $I(\vec{x})$.

Using circuit assertions, one can specify how different combinations of input and state values will determine the output and new state values. As an example, consider the 1 bit register circuit illustrated in Figure 3. We do not show any internal details of the circuit, except that the bit is stored in a feedback path involving two electrical nodes Q and $\overline{Q}$. The following assertions specify the behavior of the circuit. For readability, we depart from our vector notation and allow symbolic names for the circuit input, output, and state variables. We also allow assertions to contain variables that are implicitly universally quantified. That is, such an assertion actually represents the set of assertions obtained by replacing the free variables by all possible values they represent.

$$true\ \{Data = a \wedge Load = 1\}\ Q = a \wedge \overline{Q} = \neg a \wedge Out = a$$

$$Q = a \wedge \overline{Q} = \neg a\ \{Load = 0\}\ Q = a \wedge \overline{Q} = \neg a \wedge Out = a$$

Verifying a circuit then involves two proofs: that any circuit satisfying the set of assertions must implement the specification, and that the circuit under consideration satisfies these assertions. At present, we devise the assertions and establish their adequacy manually. This involves showing that the assertions cover every transition in the specification.

Once a set of assertions has been devised, a simulator can verify that a particular circuit satisfies them. First, we must structure the assertions in a particular way.

Definition 13: A nonvoid predicate $P\colon B^n \rightarrow \{true, false\}$ is *convex* if whenever $P(\vec{a})$ and $P(\vec{b})$ for vectors $\vec{a}$ and $\vec{b}$, then $P(\vec{c})$ for any vector $\vec{c}$ for which $c_i \in \{a_i, b_i\}$ for $1 \leq i \leq n$.

Remark: A convex predicate can be expressed by a formula of the form $P(\vec{u}) = L_1 \wedge L_2 \wedge \ldots \wedge L_k$ where each L_i is a *literal* of the form $u_j = a$ for some constant a. All the assertions for our register example our convex, and in general any assertion can be rewritten as a set of assertions containing only convex predicates.

Definition 14: For convex predicate $P(\vec{x})$, where $\vec{x} \in B^n$, the vector $\vec{x}_P \in V^n$ is defined as

$$[\vec{x}_P]_i = \begin{cases} 1, & P(\vec{a}) \Rightarrow a_i = 1 \\ 0, & P(\vec{a}) \Rightarrow a_i = 0 \\ X, & \text{else} \end{cases}$$

Remark: For a predicate expressed as a conjunction of literals, the corresponding vector is obtained by setting each element appearing in a literal to its specified value, and all other elements to X.

Theorem 2: For convex predicates P, I, N, and O if $next(\vec{z}_P, \vec{x}_I) \geq \vec{z}_N$ and $out(next(\vec{z}_P, \vec{x}_I)) \geq \vec{y}_O$ for any monotonic extension of circuit functions *next* and *out*, then the circuit satisfies assertion $P\{\ I\ \}N \wedge O$.

Proof: By monotonicity.

□

This theorem indicates a straightforward procedure to verify that a circuit satisfies an assertion by simulation. Following an Erase command, we use Set commands to set all state variables for which $[\vec{z}_P]_i \neq X$ to the appropriate values. Then we give the command $Cycle(\vec{x}_I)$ to simulate the prescribed action. Finally, we use Observe commands to check that $z_i = [\vec{z}_N]_i$ for all i such that $[\vec{z}_N]_i \neq X$, and an Output command to check that $out(\vec{z}) \geq \vec{y}_O$.

For example, to verify the register of Figure 3, we would simulate a total of 4 sequences, since there are two possible values of the free variable a.

Performance Considerations

Up to this point, we have considered only whether verifying a circuit by simulation was at all possible. Next we will investigate the practicality of this approach. In general, the circuit verification problem is NP-hard as measured in the size of the circuit and the specification. However, several techniques reduce the complexity to manageable levels for a large class of circuits.

Input Weakening

The logic value X can be used to indicate a "don't care" (or more properly "shouldn't care") condition when the circuit behavior being tested should not depend on that particular input. This allows us to simulate the effects of a number of Boolean sequences with a single ternary sequence, leading at times to a dramatic reduction in the simulation complexity. This technique is called "input weakening", because it involves reducing the information content of the simulation sequences. Monotonicity guarantees that if the resulting response on some output is 0 or 1, then all stronger sequences would give the same response.

For example, consider a k bit long, 1 bit wide shift register. The verification method described in the proof of Theorem 1 requires us to simulate 2^k patterns of the form $[a_1, a_2, \ldots, a_k]$, each time checking that $SimOut([a_1, a_2, \ldots, a_k]) = a_1$. Since the output of the shift register should depend only on the first value in the sequence, we can set the input to X for the remainder of the simulation. This reduces the number of simulation sequences to two: $[1, X, \ldots, X]$ and $[0, X, \ldots, X]$, without compromising the rigor of the simulation.

To develop this idea formally, we define a "covering set" as a set of ternary sequences that include all possible Boolean sequences of a given length.

Definition 15: A set $A \subseteq \Psi$ is a *covering set* for Φ_k if to every $\beta \in \Phi_k$ there corresponds some $\alpha \in A$ such that $\beta \geq \alpha$.

Proposition 4: For a covering set A of Φ_k, if $SimOut(\alpha) = FinalOut(q, \beta)$ for all $\alpha \in A$, all $\beta \in \Phi_k$ such that $\beta \geq \alpha$, and all $q \in Q$, then the simulated circuit $\mathcal{C}$ implements specification $\mathcal{M}$.

Input weakening can also be applied in transition simulation. In fact the simulation sequences arising from the assertion method already utilize this technique. If convex predicate P places no conditions on element i of a vector, then corresponding variable is set to X in the simulation.

Symbolic Simulation

At times we cannot avoid the complexity caused by the large number of possible input combinations that might be applied to a circuit, all of which might be relevant to the values of the outputs. For these cases, we propose *symbolic simulation* to reduce the number of patterns simulated. A symbolic simulator [3] resembles a conventional logic simulator, except that the input sequences can contain Boolean variables in addition to the constants 1 and 0. During simulation the values of the circuit state and output are Boolean functions of the variables occurring in the input sequence. A symbolic simulator represents and manipulates these functions explicitly. The worst-case behavior of such a program gives no better performance than exhaustive simulation by a conventional simulator. However, good Boolean manipulation algorithms often lead to far better results.

Discussion

Most methods of automated circuit verification utilize additional information about the circuit structure, either provided by the user or derived algorithmically. For example, the VERIFY program of Barrow [1] first constructs explicit representations of the state transition and output functions for both the circuit and the specification automata, and then attempts to prove their equivalence by algebraic manipulation. This method is prone to false negative responses (i.e. it will be unable to show the two automata are equivalent) when the structure of the specification does not closely follow that of the circuit, or when the circuit is modeled at a very detailed level.

The assumptions of this paper about the capabilities of a simulator correspond to features found in most logic simulators. Many simulators provide a logic state to represent an unknown or indeterminate value (variously denoted X, U, or $1/2$), where all state variables are assigned this value at the start of simulation [2,5]. Although an explicit Erase command may not be provided, it can easily be implemented, or the same result can be obtained by simply restarting the program. Furthermore, the ability to observe and modify selected state variables is almost universal. Requiring the simulator to compute responses that are monotonic with respect to the partial ordering $X < 1$ and $X < 0$ simply captures the desirable property that in the presence of X values, the simulator should not set an output or state variable to 0 or 1, when this would not have occurred had some of these X values been 0 or 1 instead.

We have shown that a tool normally used to test circuit designs informally can rigorously prove their correctness. However, the resulting simulation sequences differ greatly from those used by most circuit designers. Some of the ideas embodied in our simulation sequences could be applied in cases where only informal testing is desired, often leading to more reliable results at a lower cost. For example, few designers would erase the entire state of circuit except as an initial condition. However, we have shown the importance of the Erase command in detecting accidental sequential dependencies. Similarly, input weakening can often be applied to reduce the chances that the circuit behavior depends unintentionally on the value of some input. Our verification of a k bit shift register requires simulating only $2k$ patterns. In contrast, a typical designer would test a shift register by simulating some random sequence, hoping that any unintended pattern sensitivity would be detected. Both methods require approximately the same amount of work, but the first provides rigorous results while the second does not.

REFERENCES

[1] H. G. Barrow, "Proving the Correctness of Digital Hardware Designs," *VLSI Design* Vol. V, No. 7 (July, 1984) 64–77.

[2] R. E. Bryant, "A Switch-Level Model and Simulator for MOS Digital Systems," *IEEE Transactions on Computers* Vol. C-33, No. 2 (February, 1984) pp. 160–177.

[3] R. E. Bryant, "Symbolic Verification of MOS Circuits," *1985 Chapel Hill Conference on VLSI,* H. Fuchs, *ed.* Computer Science Press, 1985.

[4] L. A. Glasser, and D. W. Dobberpuhl, *The Design and Analysis of VLSI Circuits*, Addison-Wesley, 1985.

[5] J. S. Jephson, R. P. McQuarrie, and R. E. Vogelsberg, "A Three-Level Design Verification System," *IBM Systems Journal*, Vol. 8, No. 3 (1969) pp. 178–188.

[6] Z. Kohavi, *Switching and Finite Automata Theory*, McGraw-Hill, 1970.

[7] C. A. Mead, and L. Conway, *Introduction to VLSI Systems*, Addison-Wesley, 1980.

[8] E. F. Moore, "Gedanken-Experiments on Sequential Machines," *Automata Studies*, C. E. Shannon, *ed.*, Princeton University Press, 1956, pp. 129–153.

Formal Aspects of VLSI Design
G.J. Milne and P.A. Subrahmanyam (editors)

FORMAL VERIFICATION OF SYNCHRONOUS SYSTEMS

Hans Eveking

Institut für Datentechnik
Technische Hochschule Darmstadt
D-6100 Darmstadt, Fed. Rep. of Germany

A method for the formal verification of a synchronous system is presented. The method requires the specification of the system as a sequential machine and the description of the implementation as a synchronous circuit using gates and flipflops. The predicate calculus is used to show that the implementation meets the specification.

1. INTRODUCTION

The simulator-based trial-and-error approach is the most commonly used method for the functional validation of hardware descriptions. This approach has a number of well-known drawbacks:

1. test-data for hardware systems are signals, i.e., data that evolve in the time-dimension. Signals are thus more complex and more difficult to derive than the test-data for ordinary programs.
2. the simulation of a hardware system simulates the flow of signals from some fictitious environment into the system, and makes thus numerous assumptions about actual environments.
3. the simulation results must be interpreted by the designer, and must be compared with the intended function which remains informally specified ("kept in mind") in most cases.
4. simulation techniques become especially problematic if the simulation system is too weak to simulate the whole hardware system; it is extremely dubious to infer from a large number of tested parts that a system built up from these parts is correct.

In this paper, a formal verification technique will be presented which attempts to overcome the deficiencies mentioned above. The technique will be applied to the "classical" problem of the verification of a synchronous system. The intended function of the system is specified by a sequential machine while the proposed implementation is given by a synchronous circuit, i.e., a circuit that presumes a timing device like a clock-generator (see Fig. 1a and 2a for an example). The verification of the synchronous system consists of a formal proof demonstrating that the implementation meets the intended function.

2. SEQUENTIAL MACHINES AND SYNCHRONOUS CIRCUITS

2.1 An example problem

In the following, the formal verification of synchronous systems will be discussed by means of the very simple example given by Figs. 1 and 2. Fig. 1a shows a small synchronous system specified by a coded flow table (see, for example, [1]). The flow table of Fig. 1a represents the behavior of a storage element controlled by the input a: if a is L (low) then the old value of x is retained; if a is H (high) then the value of the input y is stored in x. The behavior of the sequential machine is that of a conditional transfer at the register-transfer level. In Fig. 1b, the conditional transfer is written in the computer hardware description language (CHDL) rtsmax which is a menber of the CONLAN family of CHDL's (see [2] and [3] for the details of the CONLAN approach). The conditional transfer is enclosed in the description template dsm specifying the inputs and outputs of the sequential machine.

Fig. 2a shows the proposed implementation of the sequential machine of Fig. 1 by means of a synchronous circuit consisting of a NAND-gate and a d-type flipflop. The circuit is formally represented by the description dsc of Fig. 2b. The first statement of dsc describes the NAND-gate with a delay of dnand time-units. The second statement represents the storage property of the rising-edge triggered flipflop by means of the internal carrier ff. The function re is used to detect the rising edge of the output c of the NAND-gate. The third statement reflects the propagation delay tp of the flipflop. dsc is written in the CHDL smax which is a second member of the CONLAN family of CHDL's. smax provides mechanisms for the representation of the elaborated timing details of synchronous circuits, e.g., delay times and timing conditions.

2.2 The Association of Logical Formulas with CHDL-Statements

The CHDL's rtsmax and smax are formal languages but not formal systems, i.e., we can not reason formally about the statements of Figs. 1b and 2b. We resort to the well-known formal system of the predicate calculus, and replace the CHDL-statements by logical formulas.

The association of logical formulas with the statements of CHDL's is based on the concept of time functions [4]. A time function is a mapping from the set representing the time-dimension, e.g., the set of natural numbers, in a set of values, for instance, the set {L, H} of boolean values. With each carrier of a piece of hardware, a time function is associated which represents the values that can be observed depending on the time. Fig. 3 shows examples of time functions associated with the carriers a, y and x of the sequential machine of Fig. 1.

The behavior of a hardware system represented by a set of CHDL statements is defined by a number of formulas that establish logical relationships between observations on the carriers of the hardware system. In Fig. 1c, the behavior of the sequential machine of Fig. 1a is specified by means of three formulas: (i) if the time is 0 then the value of x is L, (ii) if the time is not 0 and if the condition a at point t-1 of time is L then the old value of x is retained, (iii) if a at point t-1 of time is H then x at point t of time becomes y at point t-1 of time. Note that (ii) and (iii) correspond to the left and right half of the flow table of Fig. 1a, respectively.

Similarly, the formulas of Fig. 2c correspond to the statements of Fig. 2b.

2.3 Elements of Higher-Order Logic

The association of time functions with the carriers of a hardware system entails the introduction of certain constructs of higher-order logic. For example, the predicate re detecting the rising edge of a boolean time-function (see Fig. 2c) is specified by

$$re(F, t) \Leftrightarrow (F(t) = H) \& (F(t-1) = L).$$

In this formula, F is a function variable and t is an individual variable.

A function which takes functions as arguments is called a functional. An example of a functional is the delta-function. The delta-function was originally introduced in [5], and turned out to be one of the key concepts for the reasoning process about hardware systems. delta is defined by:

(a)

a(t-1) y(t-1)

x(t-1)	LL	LH	HH	HL
L	L	L	H	L
H	H	H	H	L

\ x(t)

(b)

```
REFLAN rtsmax
DESCRIPTION dsm(IN a, y: btml; OUT x: budvl) BODY
            IF a THEN x<- y ENDIF
ENDdsm
```

(c)

```
x(0) = L,
~(x(0) = 0)&(a(t-1) = L) => (x(t) = x(t-1)),
~(x(0) = 0)&(a(t-1) = H) => (x(t) = y(t-1))
```

Fig. 1: A sequential machine: (a) the flow table, (b) the corresponding CHDL-description enclosing a register transfer, (c) the associated axioms.

(a)

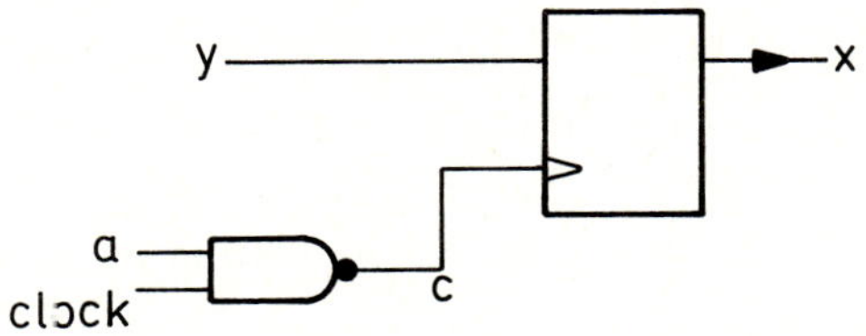

(b)

```
REFLAN smax
DESCRIPTION dsc(IN a, y, clock: btml; OUT x: btml) BODY
        DECLARE ff: bvarl; c: btml END
        c.= ~(a & clock)Δdnand,
        IF re(c) THEN ff:= y ENDIF,
        x.= ffΔtp
ENDdsc
```

(c)

```
    c(t) = not(and(a(t-dnand), clock(t-dnand))),
    ff(0) = L,
    ~(t = 0)&(re(c(t) = L) => (ff(t) = ff(t-1)),
    ~(t = 0)&(re(c(t) = H) => (ff(t) = y(t)),
    x(t) = ff(t-tp)
```

Fig. 2: A synchronous circuit: (a) the circuit diagram, (b) the corresponding CHDL description, (c) the associated axioms.

t	0	1	2	3	4	5	6	...
a(t)	L	H	L	L	H	L	H	...
y(t)	L	L	H	L	H	L	H	...
x(t)	L	L	L	L	L	H	H	...

Fig. 3: Example time functions for the sequential machine of Fig. 1

```
(delta(P, t) = z) <=> (z ≤ t) &
                      (∀ t1: (z < t1)&(t1 ≤ t) => ~P(t1)) &
                      (P(z) | (z = 0)).
```

It can be shown by induction that the delta-function satisfies the axiom:

```
(delta(P, t) = z) <=> (t = 0)&(z = 0) |
                      ~(t = 0)&P(t)&(z = t) |
                      ~(t = 0)&~P(t)&(z = delta(P, t-1)).
```

delta returns the last point of time when the predicate P was true starting with point t of time. The following example illustrates this property:

t:	0	1	2	3	4	5	6	7	8	9	...
a(t):	L	L	H	H	H	L	H	L	L	L	...
delta(λ(t1)(a(t1)=H), t):	0	0	2	3	4	4	6	6	6	6	...

delta(λ(t1)(a(t1) = H), t) returns the last point of time when the function a was high. The example shows that the introduction of anonymous functions via λ-abstraction is a second useful concept of higher-order logic.

To cope with cases where, for example, the argument P of the delta-function is delayed, a number of theorems have been derived (see APPENDIX I). An important application of the delta-function is given by the formula of Fig. 4a which can be proven to be equivalent to the formulas of Fig. 1c defining the semantics of the sequential machine of Fig. 1a. Note that in the formula of Fig. 4a, x(t) depends on a and y only and not on x itself. Hence, this formula can be used to transform statements about x into equivalent statements about a and y using the equality theorem, i.e., equals can be substituted for equals (see [6]).

Moreover, in the stationary state, i.e., for ~(delta(λ(t1)(a(t1)=H), t-1) = 0) we have simply

```
x(t) = y(delta( λ(t1)(a(t1)=H), t-1)).
```

This means that x equals the value of y at the last time when a was high. The following Fig. illustrates this property:

t:	0	1	2	3	4	5	6	7	8	9	...
a(t):	L	L	H	H	H	L	H	L	L	L	...
delta(λ(t1)(a(t1)=H), t-1):	0	0	0	2	3	4	4	6	6	6	...
y(t):	L	L	H	L	H	H	L	H	L	L	...
x(t):	L	L	L	H	L	H	H	L	L	L	...

Similarly, the formulas of Fig. 4b are equivalent to the formulas of Fig. 2c.

2.4 The Association of Theories with Hardware Descriptions

A number of concepts used in the formulas of Figs. 4a and 4b are still unspecified. For example, the meaning of terms like "t-dnand" in Fig. 4b, i.e., the meaning of the subtraction in the time dimension is not yet defined. Presuming the time to be given by the set of natural numbers, Fig. 5 formalizes the operations in the time dimension via Peano-arithmetic (see, for example, [6]). Additional axioms are necessary to define, for example, value operations like the usual boolean not-, and- and or-operation. These axioms are independently of a specific hardware description, and are called CHDL-specific axioms because they are pregiven by the particular reference language in which a hardware description is written, e.g., by the reference language rtsmax in Fig. 1b. The axioms of Fig. 4 reflecting the properties of specific descriptions, are called description-specific axioms.

```
(a)     (x(t) = z) <=> (delta( λ(t1)(a(t1)=H), t-1) = 0)&(z = L) |
                   ~(delta( λ(t1)(a(t1)=H), t-1) = 0)&
                               (z = y(delta( λ(t1)(a(t1)=H), t-1))

(b)     c(t) = not(and(a(t-dnand), clock(t-dnand))),
        (ff(t) = z) <=> (delta( λ(t1)(c(t1)=H), t) = 0)&(z = L) |
                    ~(delta( λ(t1)(c(t1)=H), t) = 0)&
                               (z = y(delta( λ(t1)(c(t1)=H), t)),
        x(t) = ff(t-tp)
```

Fig. 4: Formulas using the delta-function: (a) equivalent to 1c, (b) equivalent to 2c

```
TIME1:  ~(add1(x) = 0),
TIME2:   (add1(x) = add1(y)) => (x = y),
TIME3:   x + 0 = x,
TIME4:   x + add1(y) = add1(x + y),
TIME5:   x*0 = 0,
TIME6:   x*add1(y) = (x*y) + x,
TIME7:  ~(x < 0),
TIME8:   (x < add1(y)) <=> (x < y) | (x = y),
TIME9:   (x < y) | (x = y) | (y < x),
TIME10:  sub1(0) = 0,
TIME11:  sub1(add1(x)) = x,
TIME12: ~(x = 0) => (add1(sub1(x)) = x),
TIME13:  x - 0 = x,
TIME14:  x - add1(y) = sub1(x - y).
TIME15:  (x ≤ y) <=> (x < y) | (x = y)
```

Fig. 5: Axioms for the functions and relations in the set time

```
language L

          logical axioms
axioms <
                                   CHDL-specific axioms
          nonlogical axioms  <
                                   description-specific axioms
rules of inference
```

Fig. 6: The general structure of the theories associated with hardware descriptions

The general structure of the formal systems of the predicate calculus associated with hardware descriptions is shown in Fig. 6. The formal systems are called theories (see [6] for a survey). The description-specific and the CHDL-specific axioms define the nonlogical axioms of the theories. Since the logical axioms as well as the rules of inference are identical in all theories, specific theories are characterized by their nonlogical axioms. The axiomatization of computer hardware descriptions is discussed in detail in [7].

If a formula A of a theory T is a theorem, i.e., if it is an axiom of the theory or if it can be derived from the axioms following the rules of inference, then we write

$$|-(T)\ A.$$

We say that a statement about a hardware description is correct if it is a theorem of the associated theory.

As a result, we have obtained the formulas of Figs. 4a and 4b which define the description-specific nonlogical axioms of the sequential machine and of the synchronous circuit of Figs. 1a and 2a, respectively. These axioms plus the CHDL-specific nonlogical axioms characterize the theories associated with both descriptions.

3. RELATIONSHIPS BETWEEN HARDWARE DESCRIPTIONS AT DISTINCT LEVELS

In the last Chapter, it was demonstrated how mathematical theories can be associated with hardware descriptions. The relationships between descriptions at distinct levels of implementation, e.g., the relationship between the sequential machine and the synchronous circuit of Figs. 1 and 2, will now be discussed in terms of the relationships between the associated theories. We are particular interested in relationships which allow one to say that a high-level description is in some sense "correct" w.r.t. a low-level description. We found that proofs of correctness between descriptions at adjacent levels are that kind of "glue" which permits us to say that these descriptions represent one and the same piece of hardware.

In this Chapter, we will study only two different classes of relationships between theories by means of the example of Fig. 7 (see [8] and [9] for a more detailled discussion). Fig. 7 outlines the description of a microprogrammable machine at three distinct levels of abstraction. At levels 3 and 2, the computer is viewed as a sequential machine: at level 3, the microprogram instruction-set is specified; at level 2, the basic functional resources of the register-transfer (rt) level are represented. At level 1, the machine is viewed as a synchronous circuit, and is described as a network of integrated circuits (ic's). According to Fig. 7, we have to establish a relationship between the theories T3 and T2, and a relationship between the theories T2 and T1. For the first relationship, we will apply the concept of the extension of a theory, and will derive a formalization of the problem of logic verification (section 3.1). For the second relationship, we will use the concept of the interpretation of a theory which can be applied to more intricate problems involving temporal abstraction (section 3.2).

3.1 Logic Verification

The problem of logic verification is illustrated by Fig. 8. The levels 3 and 2 correspond to the levels of Fig. 7. At level 3, only a subset of the carriers of level 2 is represented, for example, the intervenient terminals r and s are missing at level 3. The CHDL-specific axioms of T3 and T2 coincide. We consider the description d2 to be a correct implementation of d3 if the specified logical relationships for the subset of the carriers hold in T2, i.e., if all nonlogical axioms A of T3 are theorems of T2:

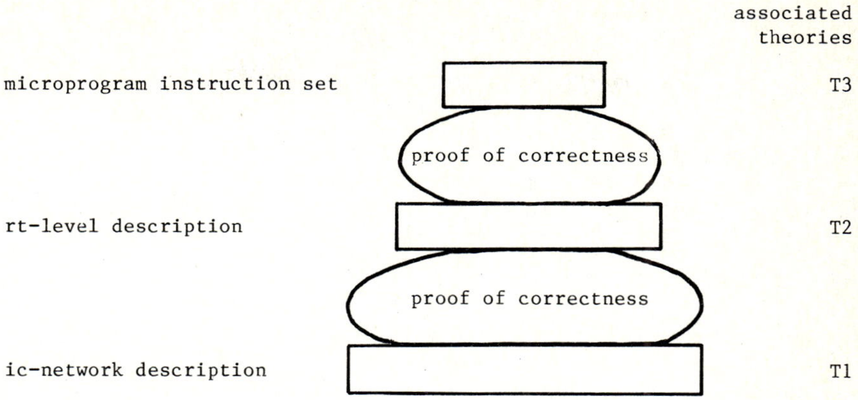

Fig. 7: Proofs of correctness are the "glue" between descriptions at distinct levels

level 3

T3: z(t) = or(and(x(t), y(t)), and(a(t), b(t))), CHDL-specific axioms

level 2

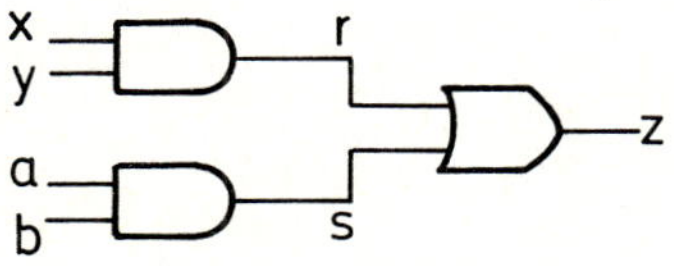

T2: z(t) = or(r(t), s(t)), r(t) = and(x(t), y(t)), s(t) = and(a(t), b(t)), CHDL-specific axioms

Fig. 8: An example of logic verification

|-(T2) A

The theory T2 is then called an extension of the theory T3 (the concept of the extension of a theory is taken from [6]).

In the example of Fig. 8, it is easily shown that T2 is an extension of T3 applying the equality theorem. In fact, propositional logic would be sufficient for this particular problem. A more complex example is given in APPENDIX II which is taken from [10] where the implementation of a logic-verifier is described. APPENDIX II shows the specification of a field of the microprogram instruction format at level 3 and the basic functional resources used at level 2 for the implementation (see also Fig. 7). The verification of this example takes about 9 seconds in interpreted INTERLISP. The logic-verifier is able to reason about the usual operations provided by register-transfer languages, e.g., vector operations like concatenation and selection, and complex boolean functions like adders and decoders.

3.2 The interpretation of theories

We will now come back to the example of Figs. 1 and 2 where a sequential machine is to be implemented by a synchronous circuit. The implementation of a sequential machine by a synchronous circuit relates the points of time at the sequential-machine level to certain points of time selected by the timing device of the synchronous circuit. In the example, the points of time are determined by the falling edge of the clock signal (see Fig. 9).

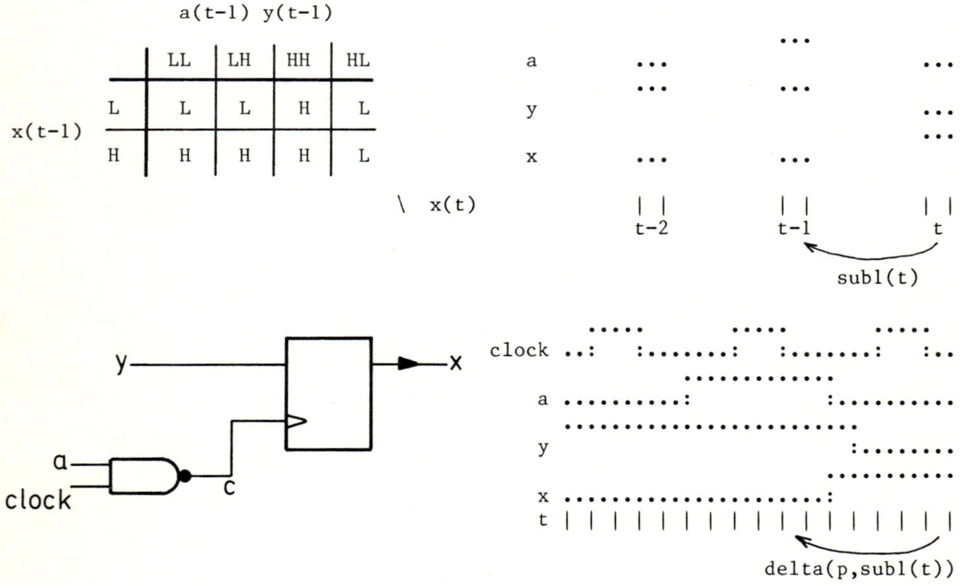

Fig. 9: The points of time of the sequential machine correspond to points of time of the synchronous circuit selected by the falling edge of the clock signal

We discuss this problem again in terms of the relationship between the associated theories T2 and T1. While the time is represented by natural numbers in T2 as well as in T1, i.e., the CHDL-specific axioms of T2 and T1 defining the time-dimension coincide, the points of time of T2, however, have a different meaning in T1. This meaning is given to T2 by an interpretation of T2 in T1. The interpretation of a theory T2 in a theory T1 means that (a) we have to define a number of rules for the translation of formulas and terms from T2 into T1, and that (b) we have to show that the interpreted axioms AI of T2 hold in T1 (see [6] for a precise definition of this concept). For our purpose, it is sufficient to define the interpretation of a theory T2 in the theory T1 by the following steps:

(1) Define a universe of interpretation by means of a predicate p of T1: Reflecting the intention of the implementation of the sequential machine, the points of time of the sequential machine correspond to those points of time of the synchronous circuit where the clock signal has a falling edge. Including t = 0, the universe of interpretation is given by the predicate:

p(t) <=> (t = 0) | (fe(clock, t) = H).

The functional fe detects the falling edge, and is defined similar to the functional re in section 2.3.

(2) Define a list of corresponding functions and predicates: For each function and predicate of T2, we define a corresponding function or predicate of T1. For instance, the sub1-function, i.e., the subtraction of 1 in T2 means that, in T1, we have to go back to the last point of time where the clock signal has a falling edge (see Fig. 9). This point of time is delta(p, sub1(t)). Hence, sub1 has λ(t)(delta(p, sub1(t))) as its corresponding function. The following list includes more examples of corresponding functions and predicates:

T2	T1
sub1	λ(t)(delta(p, sub1(t))
λ(t)(delta(q, t))	λ(t)(delta(λ(t1)(qI(t1)&p(t1)), t))
<	<

q and qI are arbitrary corresponding predicates.

(3) Show that all corresponding functions are closed: This means that the corresponding functions take values in the universe of interpretation if the arguments are in the universe. In the example of the function λ(t)(delta(p, sub1(t)) which is the function corresponding to sub1 of T2, we have to show that

|-(T1) p(t) => p(delta(p, sub1(t)))

which holds due to the definition of p and of the delta-function.

(4) The interpretation of a formula of T2 means: replace all functions and predicates by the corresponding functions and predicates. For example, if we take the formula

sub1(0) = 0

which is the axiom TIME10 of Fig. 5 and thus part of T2, then the interpreted formula is given by

delta(p, sub1(0)) = 0.

Note that this formula is an interpreted nonlogical axiom of T2.

(5) Show that for all interpreted nonlogical axioms AI of T2 holds:

```
|-(T1) p(t) => AI,
```

i.e., the interpreted nonlogical axioms hold in the universe of interpretation. For example, we have to show

```
|-(T1) p(t) => (delta(p, sub1(0)) = 0)
```

for the CHDL-specific nonlogical axiom TIME10 of T2 which is easily proven using the definition of p and delta. A more complex example is given by the interpreted description-specific nonlogical axiom of Fig. 4a:

```
|-(T1) p(t) =>
    ((x(t) = z) <=>
     (delta( λ(t1)((a(t1)=H)&(fe(clock,t1)=H)),
            delta( λ(t1)(fe(clock,t1)=H), t-1)) = 0)&(z = L) |
    ~(delta( λ(t1)((a(t1)=H)&(fe(clock,t1)=H)),
            delta( λ(t1)(fe(clock,t1)=H), t-1)) = 0)&
       (z = y(delta( λ(t1)((a(t1)=H)&(fe(clock,t1)=H)),
                   delta( λ(t1)(fe(clock,t1)=H), t-1)))))
```

where the interpretation follows from the list of corresponding functions and predicates given above. Applying the theorem DELTA5 of APPENDIX I, we have much simpler:

```
|-(T1) p(t) =>
    ((x(t) = z) <=>
     (delta( λ(t1)((a(t1)=H)&(fe(clock,t1)=H)), t-1) = 0)&(z = L) |
    ~(delta( λ(t1)((a(t1)=H)&(fe(clock,t1)=H)), t-1) = 0)&
       (z = y(delta( λ(t1)((a(t1)=H)&(fe(clock,t1)=H)), t-1)))).          (3-1)
```

3.3 Results of the proof of correctness

If we carry out the proof of 3-1, then we will find that 3-1 holds only if some assumptions about the inputs of the synchronous circuit are made. For example, the intention of the implementation is that, whenever the control signal a is high and if the clock signal has a falling edge, then the flipflop stores the value of the data-input y (see Fig. 9). However, the a and clock inputs of the NAND-gate have equal rights, i.e., if the clock signal is high and if the control signal a has a falling edge, then the flipflop is triggered, too. This undesired behavior has to be excluded by assertions on the a and clock inputs, e.g, by requiring

```
(fe(a, t) = H) => (clock(t-1) = L).
```

A similar situation is given by the implementation of a flipflop-ic by means of a gate-network. The regular flipflop-behavior does not allow for metastable states; thus, the metastable states of the gate-network are excluded by the set-up and hold time conditions specifying the assumptions on future environments of the flipflop. It is a very common pattern of hardware design that a simpler model can be obtained if restrictions on the input signals are made.

The input assertions of the synchronous circuit of Fig. 2, i.e., the conditions under which the synchronous circuit implements the sequential machine of Fig. 1 correctly are given by:

```
(fe( λ(t1)(clock(t1-dnand)), t) = H) => (stable(y, tsu, t) = H),
(fe( λ(t1)(clock(t1-dnand-th)), t) = H) => (stable(y, th, t) = H),
(fe( λ(t1)(clock(t1-dnand)), t) = H) => (stable(y, dnand, t) = H),
```

```
(fe(a, t) = H) => (clock(t-1) = L),
(fe(clock, t) = H) => (stable(a, 1, t) = H)&
   (delta( λ(t1)(fe(clock, t1)), t-1) ≤ t-tp-dnand).
```

The first two assertions are the usual set-up and hold time requirements of tsu and th time-units, transformed to the inputs of the synchronous circuit. The rest of the assertions result from the proof process. The functional stable is defined by:

```
(stable(X, ts, t) = z) <=> (∀ i: (0 ≤ i)&(i ≤ ts)&(X(t-i) = X(t)))&(z = H) |
                          ~(∀ i: (0 ≤ i)&(i ≤ ts)&(X(t-i) = X(t)))&(z = L).
```

4. THE TRANSFORMATION OF TIMING CONDITIONS

In the former Chapter, it was shown that a proof of correctness may entail a number of assertions on the inputs of a piece of hardware which future environment have to satisfy. In this Chapter, we will give a brief survey of how to prove formally the timing conditions of a piece of hardware (a detailled introduction is given in [11]).

As an example, we will discuss the transformation of a set-up time requirement made by some flipflop on its input x. The set-up time requirement is given by:

```
(re(clock, t) = H) => (stable(x, tsu, t) = H).                    (4-1)
```

Assume that the input x of the flipflop is connected with the output x of a second flipflop which is modelled by the last two statements of Fig. 2b. We ask for an assertion on the inputs y and c of the flipflop such that the assertion guarantees 4-1.

Since Fig. 4b defines the semantics of the second flipflop, we can replace x in 4-1 by λ(t2)(ff(t2-tp)) using the equality theorem. This yields:

```
(re(clock, t) = H) => (stable( λ(t2)(ff(t2-tp)), tsu, t) = H).    (4-2)
```

Moreover, due to Fig. 4b the following axiom is associated with the conditional assignment:

```
(ff(t) = z) <=> (delta( λ(t1)(c(t1)=H), t) = 0)&(z = L) |
               ~(delta( λ(t1)(c(t1)=H), t) = 0)&
                  (z = y(delta( λ(t1)(c(t1)=H), t))).
```

If we presume that the flipflop is in the stationary state, i.e., if we have ~(delta(λ(t1)(c(t1)=H), t2-tp) = 0), then

```
ff(t2-tp) = y(delta( λ(t1)(c(t1)=H), t2-tp)).
```

Using again the equality theorem, 4-2 becomes

```
(re(clock, t) = H) =>
(stable( λ(t2)(y(delta( λ(t1)(c(t1)=H), t2-tp))), tsu, t) = H).   (4-3)
```

Note that we now have transformed the original assertion 4-1 on the output x of the flipflop into an assertion on the inputs c and y. Since (t1 = t2) => (y(t1) = y(t2)) due to the uniqueness of the time function y, 4-3 is implied by

```
(re(clock, t) = H) =>
(stable( λ(t2)(delta( λ(t1)(c(t1)=H), t2-tp)), tsu, t) = H).      (4-4)
```

It is shown in [11] that 4-4 implies 4-1 for delta(λ(t1)(c(t1)=H), t2-tp) = 0, too.

The example shows that the "proof" of timing conditions results in a transformation of the assertions. We have derived procedures by which the assertions at the interface of a system can be calculated, given the internal assertions of the system [11].

5. SUMMARY

In this paper, a method for the formal verification of synchronous systems is presented. The method requires (a) the formal specification of the intended function by means of a CHDL-description at some higher level, and (b) the formal description of the implementation using a CHDL-description at a lower level. Formal systems of the predicate calculus are associated with the specification and the implementation.

If the specification abstracts from intervenient carriers only as it is the case in the problem of logic-verification, then the proof-process shows that the axioms of the specification are theorems of the theory associated with the implementation (see section 3.1 and Fig. 10).

If the specification and the implementation use different concepts of time and/or values then an appropriate interpretation of the specification-axioms must be developed to allow for a precise concept of abstraction in time, values and carriers (section 3.2). A proof of the correct implementation may result in additional input-assertions, i.e., conditions of correct implementation on the environment. This process must be repeated for all standardized modules.

For the additional input-assertions, rules of transformation must be developed. An assertions transformer is necessary which transforms internal assertions of a piece of hardware into assertions on its interface (see Fig. 10). Examples of this process are given in [9] and [12].

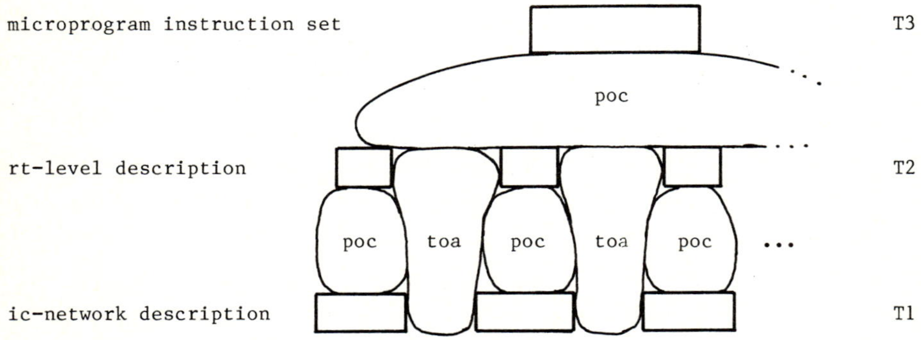

Fig. 10: The proposed verification method for the example of Fig. 7
(poc = proof of correctness, toa = transformation of assertions)

REFERENCES

[1] Hartmanis, J., Stearns, R.E., Algebraic Structure Theory of Sequential Machines, Prentice Hall, Englewood Cliffs, 1966.

[2] Piloty, R., Barbacci, M., Borrione, D., Dietmeyer, D., Hill, F. and Skelly, P., CONLAN Report, Springer-Verlag, Berlin Heidelberg New-York Tokyo, 1983.

[3] Piloty, R., Borrione, D., The CONLAN Project: Status and Future Plans, Proc. 19th Design Autom. Conf., Las Vegas, 1982, pp. 202-212.

[4] Eveking, H., Nonprocedural Specifications of Hardware, in: Teichroew, D./David, G. (Eds.), Proc. IFIP TC-2 Working Conf. on System Description Methodologies, pp. 315-332, Kecskemet (Hungary), May 1983, North-Holland.

[5] Eveking, H., The Applicaton of CONLAN ASSERTIONS to the Correct Description of Hardware, in : Breuer, M./Hartenstein, R. (Eds.), Proc. 5th Int. Symp. on CHDL, pp. 37-50, Kaiserslautern (West-Germany), Sept. 1981, North-Holland.

[6] Shoenfield, J.R., Mathematical Logic, Addison-Wesley, 1967.

[7] Eveking, H., Reasoning about Computer Hardware Descriptions, Institutsbericht RO 85/2, Institut fuer Datentechnik, Technische Hochschule Darmstadt, 1985.

[8] Eveking, H., The Application of CHDL's to the Abstract Specification of Hardware, Proc. 7th Int. Symp. on CHDL, Tokyo (Japan), Aug. 1985, North-Holland.

[9] Eveking, H., The Formal Verification of Abstract Hardware Descriptions, Institutsbericht RO 85/4, Institut fuer Datentechnik, Technische Hochschule Darmstadt, 1985.

[10] Schurig, St., Verifikation von Mikroprogrammbefehlssaetzen, Studienarbeit NVSt 273, Institut fuer Datentechnik, Technische Hochschule Darmstadt, 1985.

[11] Eveking, H., The Formal Verification of Timing Conditions, Institutsbericht RO 85/3, Institut fuer Datentechnik, Technische Hochschule Darmstadt, 1985.

[12] Eveking, H., A Structured Methodology for the Verification of Multilevel Hardware Descriptions, in: Giloi, W.K./Shriver, B.D. (Eds.), Proc. IFIP WG 10.1 Work. Conf. on Methodologies for Computer System Design, pp. 71-86, Lille (France), Sept. 1983, North-Holland.

APPENDIX I: Some important properties of the delta-function

```
DELTA1:  ~(delta(b, t) = 0) => b(delta(b, t)),

DELTA2:  delta(b, delta(b, t)) = delta(b, t),

DELTA3:  ~(t = 0) => (~(delta(b, t) = delta(b, t-1))  <=>  b(t)),

DELTA4 (delayed predicate):
        (delta( λ(t1)(b(t1-n)), t) = z) <=>
            (n ≤ t)&~(delta(b, t-n) = 0)&(z = delta(b, t-n) + n) |
            (n ≤ t)& (delta(b, t-n) = 0)& b(0)&(z = n) |
            (n ≤ t)& (delta(b, t-n) = 0)&~b(0)&(z = 0) |
           ~(n ≤ t)& b(0)&(z = t) |
           ~(n ≤ t)&~b(0)&(z = 0),

DELTA5 (conjunction of predicates):
        delta( λ(t1)(a(t1)&b(t1)), t) =
        delta( λ(t1)(a(t1)&b(t1)), delta(a, t)),

DELTA6 (disjunction of predicates):
        (delta( λ(t1)(a(t1) | b(t1)), t) = z) <=>
               (delta(a, t) < delta(b, t))&(z = delta(b, t)) |
              ~(delta(a, t) < delta(b, t))&(z = delta(a, t)),

DELTA7 (delta-of-delta):
        (delta( λ(t1)(a(delta(b, t1))), t) = z) <=>
            (delta(b, t) = 0)&(z = 0) |
           ~(delta(b, t) = 0)& a(delta(b, t))&(z = t) |
           ~(delta(b, t) = 0)&~a(delta(b, t))&
                 (z = delta( λ(t1)(a(delta(b, t1))), delta(b, t)-1)),
```

APPENDIX II: A complex example of logic verification: (a) the specification of a micro-operation (b) the implementation by functional resources corresponding to integrated circuits

(a)
```
IF (mpif[31:29] = 110)&~mpif[25]&~tfmbn&~tfmeregp&
   ~mf3&startn[3]&cpmppl THEN pc <- inc(pc) ENDIF
```

(b)
```
"/microinstruction register outputs:/"
kebefp.= mpif[31],       modrbefp.= mpif[30],
ksdbefp.= mpif[29],      popp.= mpif[25],      ...
"/microprogram sequencer: 4 AMD 2909 circuits/"
  "/microprogram counter pc:/"
IF cpmppl THEN
   IF ~tfmbn THEN pc<- inc(mbadr) ENDIF ...
ENDIF,
"/output control: oe, zero and or signals:/"
mbadr.= (source[15:8]&seq23zn  #                "/circuits 3 and 2/"
         (source[7:4] | seqlorp)&untservn  #    "/circuit 1/"
         (source[3:0] | seq0orp)&seq0zn     ),  "/circuit 0/"
"/address selection, sl and s0 signals:/"
source.= smpx(pc[15:12]#rr[15:12]#mstack[stp;0:3]#ebus[15:12],
                                              seqslp[3]#seqs0p) #
         smpx(pc[11: 8]#rr[11: 8]#mstack[stp;4:7]#ebus[11: 8],
                                              seqslp[2]#seqs0p) #
         smpx(pc[ 7: 4]#rr[ 7: 4]#mstack[stp;8:11]#ebus[ 7: 4],
                                              seqslp[1]#seqs0p) #
         smpx(pc[ 3: 0]#rr[ 3: 0]#mstack[stp;12:15]#ebus[ 3: 0],
                                              seqslp[0]#seqs0p),
IF ~tfmeregp THEN mpadr.= mbadr ENDIF,
...
"/control signals for sequencer circuits:/"
sprbedn.= ~(popp | ~bedmuxp&app[0]&sdsibefp),
sdsibefp.= ~modrbefp&~kebefp&i5sprp,
pushp.= mf3 | ~popp,
filenn.= ~mf3&~popp&untservn&~(i4upsp&seqs0p),
seqs0p.= ~sprbedn&sdsibefp,
seqslp.= smpx(sregmask[3:0] ## HHHH B,
              ksdbefp|modrbefp|kebefp|~i5sprp) & ~sprbedn,
bedmuxp.= smpx(flagr # z8ffp # z16ffp # z20ffp # flagextp # L, sb),
seq0zn.=  smpx(H # mf3n, seqmxn),
untservn.= smpx(sprbedn # H, seqmxn),
seq23zn.=  smpx(sprbedn # strtmf3n, seqmxn),
seq0orp.=  smpx(~sprbedn # ~startn[3], seqmxn),
seqmxn.= ~sdsibefp | ~i3ubrp | ~unt,
strtmf3n.= ~mf3&startn[3],
seqlorp.= ~strtmf3n, ...
```

Formal Aspects of VLSI Design
G.J. Milne and P.A. Subrahmanyam (editors)

Why higher-order logic is a good formalism for specifying and verifying hardware

Mike Gordon
Computer Laboratory
Corn Exchange Street, Cambridge CB2 3QG

Higher-order logic was originally developed as a foundation for mathematics. We show how it can be used both as a hardware description language, and as a deductive system for proving that designs meet their specifications. Examples used to illustrate various specification and verification techniques include a CMOS inverter, a CMOS full adder, an n-bit ripple-carry adder, a sequential multiplier and an edge-triggered Dtype register.

1. Introduction

The purpose of this paper is to show, via examples, that many kinds of digital systems can be formally specified using the notation of formal logic and, furthermore, that the inference rules of logic provide a practical means of proving system designs correct. We claim that there is no need for specialized hardware description languages or specialized deductive systems; 'pure logic' suffices.

The particular logical system used here is called higher-order logic. It is hoped that Section 2 below will enable readers who are not familiar with predicate calculus to understand what follows. Thorough introductions to higher-order logic can be found in textbooks on mathematical logic [8], in Church's original paper [1], or in the report on the HOL logic [4].

2. Introduction to higher-order logic

Higher-order logic uses standard predicate logic notation:

- "$P(x)$" means "x has property P",
- "$\neg t$" means "not t",
- "$t_1 \vee t_2$" means "t_1 or t_2",
- "$t_1 \wedge t_2$" means "t_1 and t_2",
- "$t_1 \supset t_2$" means "t_1 implies t_2",

- "$t_1 \equiv t_2$" means "t_1 if and only if t_2",
- "$\forall x.\ t[x]$" means "for all x it is the case that $t[x]$",
- "$\exists x.\ t[x]$" means "for some x it is the case that $t[x]$",
- "$(t \rightarrow t_1 \mid t_2)$" means "if t is true then t_1 else t_2".

There are three important ways that higher-order logic extends first-order logic.

1. Variables can range over functions and predicates. Such variables are called *higher-order* and can be quantified. For example, the principle of mathematical induction can be expressed using a variable P that ranges over predicates:

$$\forall P.\ P(0) \wedge (\forall n.\ P(n) \supset P(n{+}1)) \supset \forall n.\ P(n)$$

The existence of a function satisfying a simple recursive definition can be stated using variables f and s that range over functions.

$$\forall n_0.\ \forall f.\ \exists s.\ (s(0) = n_0) \wedge (\forall n.\ s(n{+}1) = f(s(n)))$$

This asserts that for each number n_0 and function f there exists a function s such that $s(0){=}n_0$ and $s(n{+}1){=}f(s(n))$ for all n. The two examples just described make essential use of higher-order variables, and thus they can't be expressed in first-order logic.

2. Functions and predicates can be the arguments and results of other functions and predicates. For example, a function Iterate can be defined such that:

$$\mathsf{Iterate}(m,n)(f) \ \equiv\ f(n) \ \wedge\ f(n{-}1) \ \wedge\ \cdots\ \wedge\ f(m)$$

Iterate maps a pair of numbers (m, n) to a higher-order predicate whose argument is a function. This predicate is true of a function f if and only if $f(n) \wedge f(n{-}1) \wedge \cdots \wedge f(m)$ holds.

3. Higher-order logic has special function-denoting terms called λ-expressions. These have the form $\lambda x.\ t$ where x is a variable and t is an expression. Such a λ-term denotes the function $a \mapsto t[a/x]$ where $t[a/x]$ is the result of substituting a for x in t. For example, $\lambda x.\ x{+}3$ denotes the function $a \mapsto a{+}3$ which adds 3 to its argument. A common use of λ-expressions is as arguments to higher-order functions. It is explained in Section 7.2.2 below how the term

$$\mathsf{Iterate}\ (0, n)\ (\lambda i.\ \mathsf{Add1}(a(i), b(i), c(i), sum(i), c(i{+}1)))$$

can be used to represent an array of $n{+}1$ 1-bit adders.

3. Representing behaviour with predicates

A device is a 'black box' with a specified behaviour; for example:

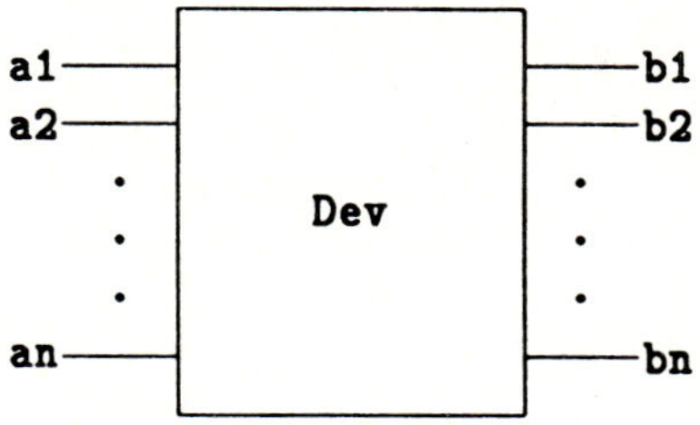

This device is called `Dev` and has external lines `a1`, `a2`, ... , `am`, `b1`, `b2`, ... , `bn`. These lines correspond to the 'pins' of an integrated circuit. When the device is in operation each line has a value drawn from some set of possible values. Different kinds of device are modelled with different sets of values. The behaviour of device `Dev` is specified by defining a predicate Dev (with $m+n$ arguments) such that $\mathsf{Dev}(a_1, a_2, \ldots, a_m, b_1, b_2, \ldots, b_n)$ holds if and only if a_1, a_2, ... , a_m, b_1, b_2, ... , b_n are allowable values on the corresponding lines of `Dev`.

The following font conventions will be used:

- Physical objects like devices and lines will be written in `typewriter` font.
- Mathematical variables will be written in *italic* font.
- Mathematical constants, (*e.g.* predicate and function constants) will be written in sans serif font.

The same letter will be used for a physical object and its mathematical representation. Thus, for example, l will range over the values allowed at line `l`, and Dev denotes the predicate describing the behaviour of device `Dev`.

We now describe two examples that illustrate the use of predicates to specify behaviour. In the first of these examples the values on lines are modelled with truth-values. In the second example the values on lines are modelled with functions, and consequently the predicate used to specify the behaviour of the device is higher-order.

3.1. A delayless switch

Zero-delay combinational devices can be modelled by taking the boolean values T and F as the allowed values on their lines. An example is a switch:

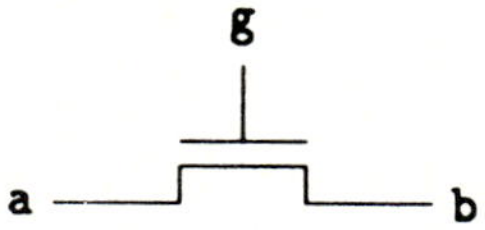

The intended behaviour of this is that **a** is connected to **b** if **g** has the value T and **a** and **b** are not connected if **g** has the value F. This behaviour can be represented by the predicate Switch defined by:

$$\mathsf{Switch}(g, a, b) \;\equiv\; (g \supset (a = b))$$

The condition $\mathsf{Switch}(g, a, b)$ holds if and only if whenever g is true then a and b are equal. For example, $\mathsf{Switch}(\mathsf{T}, \mathsf{F}, \mathsf{F})$ holds because $\mathsf{T} \supset (\mathsf{F}{=}\mathsf{F})$ is true, and $\mathsf{Switch}(\mathsf{F}, \mathsf{T}, \mathsf{F})$ holds because $\mathsf{F} \supset (\mathsf{T}{=}\mathsf{F})$ is true, but $\mathsf{Switch}(\mathsf{T}, \mathsf{T}, \mathsf{F})$ does not hold because $\mathsf{T} \supset (\mathsf{T}{=}\mathsf{F})$ is false.

3.2. An inverter with delay

The values occurring on the lines of devices may vary over time. When this happens, their behaviour must be represented by predicates whose arguments are 'time varying values'. Such values correspond to 'waveforms' and can be modelled by functions of time. For example, the behaviour of an inverter with a delay of δ units of time can be specified with a predicate Invert defined by:

$$\mathsf{Invert}(i, o) \;\equiv\; \forall t.\ o(t{+}\delta) = \neg i(t)$$

Here the values on lines **i** and **o** are functions i and o which map times (represented by numbers) to values (represented by booleans). These functions are in the Invert relation if and only if for all times t, the value of o at time $t{+}\delta$ equals the value of i at time t.

4. Representing circuit structure with predicates

Consider the following structure (called D):

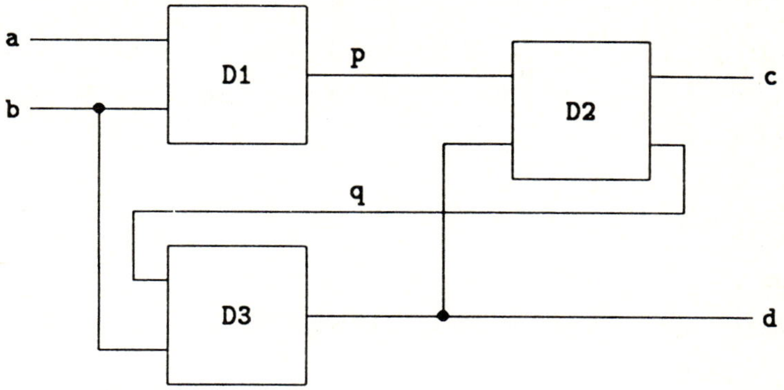

This device is built by connecting together three component devices D1, D2 and D3. The external lines of D are a, b, c and d. The lines p and q are internal and are not connected to the 'outside world'. (External lines might correspond to the pins of an integrated circuit, and internal lines to tracks.)

Suppose the behaviours of D1, D2 and D3 are specified by predicates D_1, D_2 and D_3 respectively. How can we derive the behaviour of the system D shown above? Each device constrains the values on its lines. If a, b and p denote the values on the lines a, b and p, then D1 constrains these values so that $D_1(a,b,p)$ holds. To get the constraint imposed by the whole device D we just conjoin (*i.e.* $\wedge$-together) the constraints imposed by D1, D2 and D3; the combined constraint is thus:

$$D_1(a,b,p) \;\wedge\; D_2(p,d,c) \;\wedge\; D_3(q,b,d)$$

This expression constrains the values on both the external lines a, b, c and d and the internal lines p and q. If we regard D as a 'black box' with the internal lines invisible, then we are really only interested in what constraints are imposed on its external lines. The variables a, b, c and d will denote possible values at the external lines a, b, c and d if and only if the conjunction above holds *for some* values p and q. We can therefore define a predicate D representing the behaviour of D by:

$$D(a,b,c,d) \;\equiv\; \exists p\ q.\ D_1(a,b,p) \;\wedge\; D_2(p,d,c) \;\wedge\; D_3(q,b,d)$$

Thus we see that the behaviour corresponding to a circuit is got by:

- Conjoining the constraints corresponding to the components, and
- existentially quantifying the variables corresponding to the internal lines.

This technique of representing circuit diagrams in logic is fairly well known [9]. In the terminology of CCS [11] we are using conjunction for parallel composition and existential quantification for hiding. Other ways of representing structure in logic are also possible [2].

5. A CMOS inverter

The standard CMOS implementation of an inverter is:

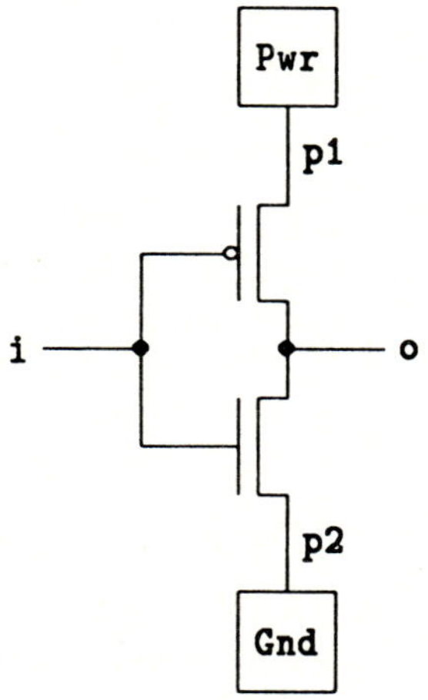

5.1. Specification of the components

The inverter shown above can be viewed as a structure built out of four components: a power source, a ground, an n-transistor and a p-transistor.

5.1.1. Power

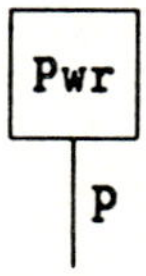

This is a power source (sometimes called V_{dd}) and can be modelled by a predicate

Pwr that constrains the value on the line **p** to be T.

$$\mathsf{Pwr}(p) \equiv (p = \mathsf{T})$$

5.1.2. Ground

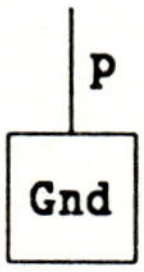

This represents 'ground' and can be modelled by a predicate Gnd that constrains the value on the line **p** to be F.

$$\mathsf{Gnd}(p) \equiv (p = \mathsf{F})$$

5.1.3. *n*-transistor

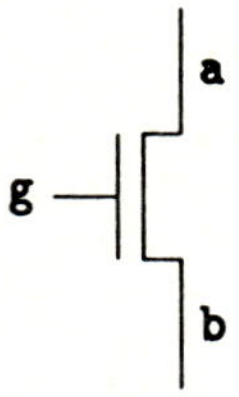

This represents an *n*-transistor. It can be modelled as a switch.

$$\mathsf{Ntran}(g, a, b) \equiv (g \supset (a = b))$$

5.1.4. *p*-transistor

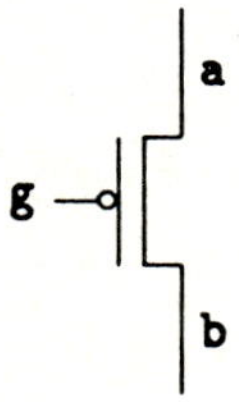

This represents a *p*-transistor. It can be modelled as a switch which conducts when its gate (*i.e.* line **g**) is low.

$$\mathsf{Ptran}(g, a, b) \equiv (\neg g \supset (a = b))$$

5.2. Logic representation of the inverter circuit

Conjoining together the constraints from the four components and existentially quantifying the internal line variables yields the following definition of a predicate Inv:

$$\mathsf{Inv}(i,o) \;\equiv\; \exists p_1\ p_2.\ \mathsf{Pwr}(p_1) \wedge \mathsf{Ptran}(i,p_1,o) \wedge \mathsf{Ntran}(i,o,p_2) \wedge \mathsf{Gnd}(p_2)$$

If $\mathsf{Inv}(i,o)$ holds then the values i and o are contrained to be in the relation determined by the inverter circuit above.

5.3. Verification by proof

It follows by standard logical reasoning that if Inv is defined as above, then

$$\mathsf{Inv}(i,o) \;\equiv\; (o = \neg i)$$

This shows that the constraint on i and o imposed by the inverter circuit is exactly what we want: o is the inverse of i.

An outline of the formal proof of this is as follows:

1. By definition of Inv:

$$\begin{aligned}\mathsf{Inv}(i,o) \;\equiv\; \exists p_1\ p_2.\ & \mathsf{Pwr}(p_1) \wedge \\ & \mathsf{Ptran}(i,p_1,o) \wedge \\ & \mathsf{Ntran}(i,o,p_2) \wedge \\ & \mathsf{Gnd}(p_2)\end{aligned}$$

2. Substituting in the definitions of Pwr and Gnd yields:

$$\begin{aligned}\mathsf{Inv}(i,o) \;\equiv\; \exists p_1\ p_2.\ & (p_1 = \mathsf{T}) \wedge \\ & \mathsf{Ptran}(i,p_1,o) \wedge \\ & \mathsf{Ntran}(i,o,p_2) \wedge \\ & (p_2 = \mathsf{F})\end{aligned}$$

3. Substituting with the equations $p_1 = \mathsf{T}$ and $p_2 = \mathsf{F}$ yields:

$$\begin{aligned}\mathsf{Inv}(i,o) \;\equiv\; \exists p_1\ p_2.\ & (p_1 = \mathsf{T}) \wedge \\ & \mathsf{Ptran}(i,\mathsf{T},o) \wedge \\ & \mathsf{Ntran}(i,o,\mathsf{F}) \wedge \\ & (p_2 = \mathsf{F})\end{aligned}$$

4. In general, if t_1 and t_2 are any terms such that t_2 doesn't contain x then:

$$(\exists x.\ t_1 \ \wedge\ t_2) \ = ((\exists x.\ t_1) \ \wedge\ t_2)$$

and

$$(\exists x.\ t_2 \ \wedge\ t_1) \ = (t_2 \ \wedge\ (\exists x.\ t_1))$$

are both true. Using these properties we can move the existential quantifiers inwards to derive:

$$\begin{aligned}\mathsf{Inv}(i,o) \ \equiv\ & (\exists p_1.\ p_1 = \mathsf{T}) \wedge \\ & \mathsf{Ptran}(i,\mathsf{T},o) \wedge \\ & \mathsf{Ntran}(i,o,\mathsf{F}) \wedge \\ & (\exists p_2.\ p_2 = \mathsf{F})\end{aligned}$$

5. Both $(\exists p_1.\ p_1{=}\mathsf{T})$ and $(\exists p_2.\ p_2{=}\mathsf{F})$ are logical truths and so can be deleted from conjunctions; hence:

$$\mathsf{Inv}(i,o) \ \equiv\ \mathsf{Ptran}(i,\mathsf{T},o) \ \wedge\ \mathsf{Ntran}(i,o,\mathsf{F})$$

6. Next we substitute in the definitions of **Ptran** and **Ntran** to get:

$$\mathsf{Inv}(i,o) \ \equiv\ ((i = \mathsf{F}) \supset (\mathsf{T} = o)) \ \wedge\ ((i = \mathsf{T}) \supset (o = \mathsf{F}))$$

7. From this we can derive

$$\mathsf{Inv}(\mathsf{T},o) \ \equiv\ (o = \mathsf{F})$$

and

$$\mathsf{Inv}(\mathsf{F},o) \ \equiv\ (o = \mathsf{T})$$

from which it follows by case analysis that:

$$\mathsf{Inv}(i,o) \ \equiv\ (o = \neg i)$$

Proofs such as these can be generated by the HOL system [5].

6. A 1-bit CMOS full adder

The full adder described in this section was shown to me by Inder Dhingra; it illustrates the use of bidirectional transistors in CMOS. The transistor models Ptran and Ntran can be used to prove the circuit correct. Such a proof would be difficult with the usual representation of combinational circuits as boolean functions. Relations rather than functions are needed to model bidirectionality.

6.1. Specification

Here is a diagram of a full adder:

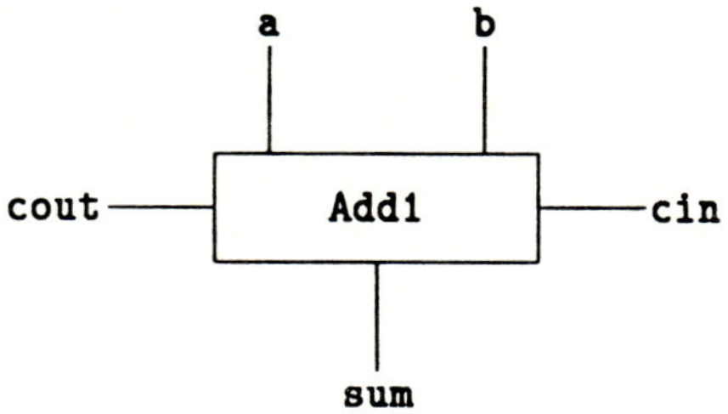

The lines a, b, cin, sum and cout carry the boolean values T or F. To relate these logical values to the numbers 1 and 0 we define a function Bit_Val by:

$$\mathsf{Bit\_Val}(\mathsf{T}) = 1 \quad \text{and} \quad \mathsf{Bit\_Val}(\mathsf{F}) = 0$$

The specification of the adder can then be defined by:

$$\begin{aligned}
&\mathsf{Add1}(a, b, cin, sum, cout) \;\equiv \\
&\;(2 \times \mathsf{Bit_Val}(cout) + \mathsf{Bit_Val}(sum) = \\
&\;\;\mathsf{Bit_Val}(a) + \mathsf{Bit_Val}(b) + \mathsf{Bit_Val}(cin))
\end{aligned}$$

A correct implementation of this specification is a circuit with lines a, b, cin, sum and cout such that the constraints imposed on the values a, b, cin, sum and, $cout$ that can exist on these lines imply that $\mathsf{Add1}(a, b, cin, sum, cout)$ always holds.

6.2. Implementation

A CMOS implementation of the adder is given below. Lines with the same name are connected. The lines p0, ... , p11 are internal. The two transistors drawn horizontally function bidirectionally.

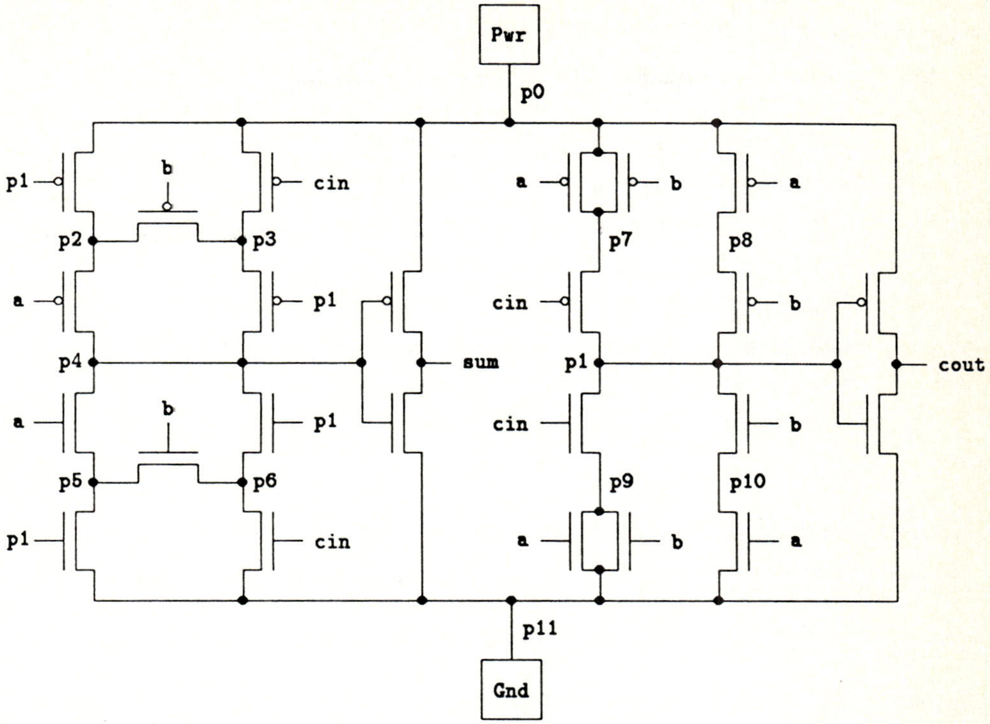

This circuit can be represented in logic by defining:

$$\begin{array}{l}
\textsf{Add1_Imp}(a, b, cin, sum, cout) \ \equiv \\
\exists p_0\ p_1\ p_2\ p_3\ p_4\ p_5\ p_6\ p_7\ p_8\ p_9\ p_{10}\ p_{11}. \\
\quad \textsf{Ptran}(p_1, p_0, p_2) \wedge \textsf{Ptran}(cin, p_0, p_3) \wedge \textsf{Ptran}(b, p_2, p_3) \wedge \textsf{Ptran}(a, p_2, p_4) \wedge \\
\quad \textsf{Ptran}(p_1, p_3, p_4) \wedge \textsf{Ntran}(a, p_4, p_5) \wedge \textsf{Ntran}(p_1, p_4, p_6) \wedge \textsf{Ntran}(b, p_5, p_6) \wedge \\
\quad \textsf{Ntran}(p_1, p_5, p_{11}) \wedge \textsf{Ntran}(cin, p_6, p_{11}) \wedge \textsf{Ptran}(a, p_0, p_7) \wedge \textsf{Ptran}(b, p_0, p_7) \wedge \\
\quad \textsf{Ptran}(a, p_0, p_8) \wedge \textsf{Ptran}(cin, p_7, p_1) \wedge \textsf{Ptran}(b, p_8, p_1) \wedge \textsf{Ntran}(cin, p_1, p_9) \wedge \\
\quad \textsf{Ntran}(b, p_1, p_{10}) \wedge \textsf{Ntran}(a, p_9, p_{11}) \wedge \textsf{Ntran}(b, p_9, p_{11}) \wedge \textsf{Ntran}(a, p_{10}, p_{11}) \wedge \\
\quad \textsf{Pwr}(p_0) \wedge \textsf{Ptran}(p_4, p_0, sum) \wedge \textsf{Ntran}(p_4, sum, p_{11}) \wedge \\
\quad \textsf{Gnd}(p_{11}) \wedge \textsf{Ptran}(p_1, p_0, cout) \wedge \textsf{Ntran}(p_1, cout, p_{11})
\end{array}$$

6.3. Verification

To verify that the implementation **Add1_Imp** correctly implements the specification **Add1**, it must be proved that if a, b, cin, sum and $cout$ satisfy the constraints imposed by the implementation, then they also satisfy the specification. Formally:

$$\textsf{Add1\_Imp}(a, b, cin, sum, cout) \ \supset \ \textsf{Add1}(a, b, cin, sum, cout)$$

The most straightforward way to prove this implication is to consider separately the eight possible input combinations.

Manipulations like those given for the inverter yield the following eight facts:

$$\begin{aligned}
\mathsf{Add1_Imp}(\mathsf{T},\mathsf{T},\mathsf{T},sum,cout) &\equiv (sum{=}\mathsf{T}) \wedge (cout{=}\mathsf{T})\\
\mathsf{Add1_Imp}(\mathsf{T},\mathsf{T},\mathsf{F},sum,cout) &\equiv (sum{=}\mathsf{F}) \wedge (cout{=}\mathsf{T})\\
\mathsf{Add1_Imp}(\mathsf{T},\mathsf{F},\mathsf{T},sum,cout) &\equiv (sum{=}\mathsf{F}) \wedge (cout{=}\mathsf{T})\\
\mathsf{Add1_Imp}(\mathsf{T},\mathsf{F},\mathsf{F},sum,cout) &\equiv (sum{=}\mathsf{T}) \wedge (cout{=}\mathsf{F})\\
\mathsf{Add1_Imp}(\mathsf{F},\mathsf{T},\mathsf{T},sum,cout) &\equiv (sum{=}\mathsf{F}) \wedge (cout{=}\mathsf{T})\\
\mathsf{Add1_Imp}(\mathsf{F},\mathsf{T},\mathsf{F},sum,cout) &\equiv (sum{=}\mathsf{T}) \wedge (cout{=}\mathsf{F})\\
\mathsf{Add1_Imp}(\mathsf{F},\mathsf{F},\mathsf{T},sum,cout) &\equiv (sum{=}\mathsf{T}) \wedge (cout{=}\mathsf{F})\\
\mathsf{Add1_Imp}(\mathsf{F},\mathsf{F},\mathsf{F},sum,cout) &\equiv (sum{=}\mathsf{F}) \wedge (cout{=}\mathsf{F})
\end{aligned}$$

Deriving these equations is equivalent to exhaustive simulation for all input values and is best done by computer. It follows from these eight equations that:

$$\mathsf{Add1\_Imp}(a,b,cin,sum,cout) \equiv \mathsf{Add1}(a,b,cin,sum,cout)$$

This entails the implication we want, since if $t_1{\equiv}t_2$ then $t_1{\supset}t_2$ holds *a fortiori*.

7. An n-bit adder

The example in this section is intended to illustrate the use of higher-order logic to represent parameterized systems. An n-bit adder computes an n-bit sum and 1-bit carry-out from two n-bit inputs and a 1-bit carry-in. Each value of the parameter n determines an adder operating on words of size $n+1$ ("$n+1$" because n counts from 0; *e.g.* if n is 0 we get a 1-bit adder).

7.1. Specification

Here is a diagram of an n-bit adder:

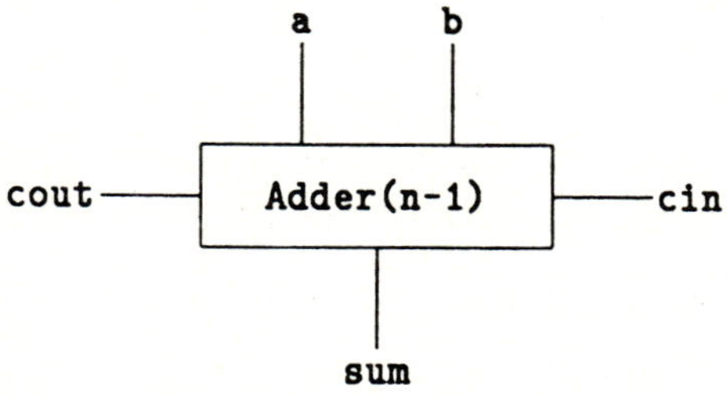

The lines **cin** and **cout** carry 1-bit words and the lines **a**, **b** and **sum** carry n-bit words. We will model 1-bit words as booleans and n-bit words as functions from natural numbers to booleans. Thus the 4-bit word 1101 is represented by a function f such that $f(0)=\mathsf{T}$, $f(1)=\mathsf{F}$, $f(2)=\mathsf{T}$ and $f(3)=\mathsf{T}$. To relate words to numbers we use a function Val such that $\mathsf{Val}(n, f)$ is the number denoted by the $(n+1)$-bit word $f(n)f(n-1)\cdots f(0)$. Val can be defined by ***primitive recursion.***

The primitive recursive definition of Val consists of two parts, a ***basis*** which specifies $\mathsf{Val}(0, f)$ and a ***recursion*** which specifies $\mathsf{Val}(n+1, f)$ in terms of $\mathsf{Val}(n, f)$. The basis is:

$$\mathsf{Val}(0, f) = \mathsf{Bit\_Val}(f(0))$$

and the recursion is:

$$\mathsf{Val}(n+1, f) = 2^{n+1} \times \mathsf{Bit\_Val}(f(n+1)) + \mathsf{Val}(n, f)$$

To specify an n-bit adder we will define a higher-order function Adder which when applied to the number $n-1$ yields a predicate specifying the adder. Thus, for example, Adder(3) is a predicate specifying a 4-bit adder. The definition of Adder is:

$$\begin{aligned}&\mathsf{Adder}(n)(a, b, cin, sum, cout) \ \equiv \\ &(2^{n+1} \times \mathsf{Bit\_Val}(cout) + \mathsf{Val}(n, sum) = \\ &\ \mathsf{Val}(n, a) + \mathsf{Val}(n, b) + \mathsf{Bit\_Val}(cin))\end{aligned}$$

7.2. Implementation

An n-bit adder can be built by connecting together n full adders. The diagram below shows an $(n+1)$-bit adder. The inputs are a single bit carry-in **cin** and two $(n+1)$-bit words **a(n)a(n-1)...a(0)** and **b(n)b(n-1)...b(0)**. The outputs are an $(n+1)$-bit sum **sum(n)sum(n-1)...sum(0)** and a 1-bit carry-out **cout**.

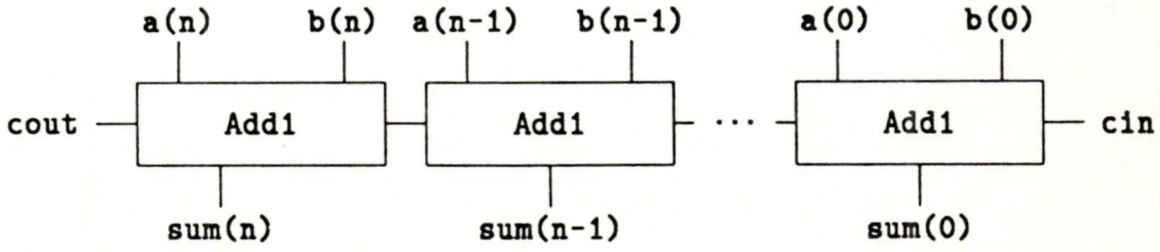

To express this diagram in logic we define $\mathsf{Adder\_Imp}(n)(a, b, cin, out, cout)$ where $\mathsf{Adder\_Imp}$ is a higher-order function which when applied to a number n yields the predicate specifying the implementation of an $n+1$-bit adder.

Two logically equivalent definitions of $\mathsf{Adder\_Imp}$ are given below. The first one is a simple primitive recursive definition. The second one is intended to mimic how the adder might be described in a commercial hardware description language like MODEL [10]. This second definition illustrates the claim that pure logic has all the expressive power found in special purpose languages.

7.2.1. Recursive description of the adder circuit

A primitive-recursive definition of $\mathsf{Adder\_Imp}$ corresponding to the above diagram has the following basis:

$$\mathsf{Adder\_Imp}(0)(a, b, cin, sum, cout) \equiv \mathsf{Add1}(a(0), b(0), cin, sum(0), cout)$$

The recursive part of the definition says that an $(n+2)$-bit adder is built by first building an $n+1$-bit adder and then connecting its carry-out to the carry-in of a 1-bit adder.

$$\begin{array}{l} \mathsf{Adder\_Imp}(n+1)(a, b, cin, sum, cout) \equiv \\ \quad \exists c.\ \mathsf{Adder\_Imp}(n)(a, b, cin, sum, c) \wedge \\ \qquad \mathsf{Add1}(a(n+1), b(n+1), c, sum(n+1), cout) \end{array}$$

7.2.2. Iterative description of the adder circuit

To indicate the expressiveness of higher-order logic we will show how to mimic the following 'part specification' written in MODEL.

```
PART Adder_Imp (n) [a(0:n),b(0:n),cin] -> sum(0:n),cout
 SIGNAL c(0:n+1)
 INTEGER i
 cin -> c(0)
 FOR i = 0:n CYCLE
  Add1[a(i),b(i),c(i)] -> sum(i),c(i+1)
 REPEAT
 c(n+1) -> cout
END
```

Here is a line by line explanation of this code.

1. `PART Adder_Imp (n) [a(0:n),b(0:n),cin] -> sum(0:n),cout`

   A part called `Adder_Imp` is being specified.

   (a) `(n)`

   indicates that the part is parameterized on a number `n`.

   (b) `[a(0:n),b(0:n),cin]`

   indicates that `Adder_Imp` has two `(n+1)`-bit inputs called `a` and `b`, and one single bit input called `cin`.

   (c) `-> sum(0:n),cout`

   indicates that `Adder_Imp` has an `(n+1)`-bit output called `out` and a 1-bit output called `cout`.

2. `SIGNAL c(0:n+1)`

   declares a 'local' `(n+1)`-bit line called `c`.

3. `INTEGER i`

   declares a local integer-valued variable called `i`.

4. `cin -> c(0)`

   specifies that `cin` be connected to the `0`th bit of `c`.

5. `FOR i = 0:n CYCLE`

   starts an iteration in which the body of the iteration (see below) is executed with `i` successively taking values `0`, `1`, ... , `n`.

6. `Add1[a(i),b(i),c(i)] -> sum(i),c(i+1)`

   is the body of the iteration; it specifies an instance of `Add1` having as inputs the `i`th bits of `a`, `b` and `c` and as outputs the `i`th bit of `sum` and the `i+1`th bit of `c`.

7. `REPEAT`

   ends the iteration.

8. `c(n+1) -> cout`

   specifies that the `n+1`th bit of `c` be connected to `cout`.

9. `END`

   ends the specification of `Adder_Imp`.

A transcription of the MODEL part specification into higher-order logic is:

$$
\begin{aligned}
&\mathsf{Adder\_Imp}(n)(a, b, cin, sum, cout) \equiv \\
&\exists c. \\
&\ \ cin = c(0) \ \wedge \\
&\ \ \mathsf{Iterate}\ (0, n)\ (\lambda i.\ \mathsf{Add1}(a(i), b(i), c(i), sum(i), c(i+1))) \ \wedge \\
&\ \ c(n+1) = cout
\end{aligned}
$$

Note that in logic:

- a 'part specification' is just the definition of a function,
- a 'signal declaration' is an existential quantification,
- a wiring specification (*e.g.* `cin -> c(0)`) is an equation, and
- an iteration is just an invocation of the higher-order function Iterate.

No *ad hoc* hardware description constructs are needed, pure logic is enough.

7.3. Verification

To verify the adder one proves by induction on n that:

$$\mathsf{Adder\_Imp}(n)(a, b, cin, sum, cout) \supset \mathsf{Adder}(n)(a, b, cin, sum, cout)$$

The basis of the induction is:

$$\mathsf{Adder\_Imp}(0)(a, b, cin, sum, cout) \supset \mathsf{Adder}(0)(a, b, cin, sum, cout)$$

This is easily proved by substituting the definitions of Adder_Imp and Adder into the above implication and then reducing the resulting expression to a tautology. The induction step is:

$$
\begin{aligned}
&(\mathsf{Adder\_Imp}(n)(a, b, cin, sum, cout) \supset \mathsf{Adder}(n)(a, b, cin, sum, cout)) \\
&\supset \\
&(\mathsf{Adder\_Imp}(n+1)(a, b, cin, sum, cout) \supset \mathsf{Adder}(n+1)(a, b, cin, sum, cout))
\end{aligned}
$$

This can be proved by simple arithmetic. A correctness proof of the recursively specified adder has been generated by Albert Camilleri using the HOL system.

8. Sequential Devices

All of the examples so far have been combinational; *i.e.* the values on the outputs have only depended on the current input values, not on input values at past times. Sequential devices can be modelled by taking the values on lines to be functions of time. For example, a unit-delay element `Del`, with input line `i` and output line `o`, is modelled by specifying that the value output at time $t+1$ equals the value input at time t. This is expressed in higher-order logic by:

$$\mathsf{Del}(i, o) \;\equiv\; \forall t.\; o(t+1) = i(t)$$

Combinational devices can be modelled as sequential devices having no delay. To illustrate this, recall the specification of the adder:

$$\begin{array}{l} \mathsf{Adder}(n)(a, b, cin, sum, cout) \;\equiv \\ \;(2^{n+1} \times \mathsf{Bit\_Val}(cout) + \mathsf{Val}(n, sum) = \\ \;\;\mathsf{Val}(n, a) + \mathsf{Val}(n, b) + \mathsf{Bit\_Val}(cin)) \end{array}$$

The variables a, b and sum range over words (modelled as functions) and the variables cin and $cout$ range over truth-values. To model the adder as a zero-delay sequential device we must represent its behaviour with a predicate whose arguments are functions of time.

$$\begin{array}{l} \mathsf{Combinational\_Adder}(n)(a', b', cin', sum', cout') \;\equiv \\ \;\forall t.\; \mathsf{Adder}(n)(a'(t), b'(t), cin'(t), sum'(t), cout'(t)) \end{array}$$

The variables a', b' and sum' range over functions from time to words, and the variables cin' and $cout'$ range over functions from time to truth-values. Thus, for example, $out'(7)(5)$ is bit 5 of the word output at time 7. If we wanted to specify the adder as having a unit-delay then we could define:

$$\begin{array}{l} \mathsf{Unitdelay\_Adder}(n)(a', b', cin', sum', cout') \;\equiv \\ \;\forall t.\; \mathsf{Adder}(n)(a'(t), b'(t), cin'(t), sum'(t+1), cout'(t+1)) \end{array}$$

9. A sequential multiplier

As an illustration of the specification and verification of a sequential device, we describe below a multiplier. To simplify details, it is (unrealistically) assumed that lines carry numbers rather than words. This enables us to use ordinary arithmetic operators in the specification.

9.1. Specification

The multiplier is specified to have two inputs and two outputs.

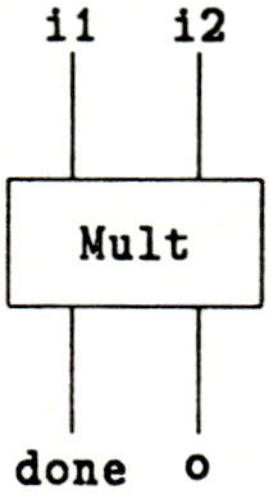

An informal specification of the required behaviour of `Mult` is:

If

- **done** has value T at time t_1, and
- t_2 is the first time after t_1 that **done** again has value T, and
- the values at `i1` and `i2` are stable from t_1 to t_2,

then

- the value at **o** at time t_2 is the product of the values at `i1` and `i2` at time t_1.

In order to formalize this in logic various temporal notions like "the first time after" and "stable" must be represented.

9.1.1. Some temporal predicates

The predicate Stable is defined so that $\mathsf{Stable}(t_1, t_2)(f)$ is true if and only if the value of f is constant from t_1 until just before time t_2. Formally:

$$\mathsf{Stable}(t_1,t_2)(f) \;\equiv\; \forall t.\; t_1 \leq t \;\wedge\; t < t_2 \;\supset\; (f(t)=f(t_1))$$

The predicate Next is defined so that $\mathsf{Next}(t_1, t_2)(f)$ is true if and only if t_2 is the first time after t_1 that $f(t_2)=\mathsf{T}$. Formally:

$$\mathsf{Next}(t_1,t_2)(f) \;\equiv\; t_1 < t_2 \;\wedge\; f(t_2) \;\wedge\; (\forall t.\; t_1 < t \;\wedge\; t < t_2 \;\supset\; \neg f(t))$$

Using **Stable** and **Next**, the specification of `Mult` can be represented with the predicate **Mult** defined by:

$$\begin{aligned}&\mathsf{Mult}(i_1, i_2, o, done) \equiv \\ &done(t_1) \wedge \\ &\mathsf{Next}(t_1, t_2)(done) \wedge \\ &\mathsf{Stable}(t_1, t_2)(i_1) \wedge \\ &\mathsf{Stable}(t_1, t_2)(i_2) \wedge \\ &\supset \\ &(o(t_2) = i_1(t_1) \times i_2(t_1))\end{aligned}$$

9.2. Implementation

Here is a circuit which meets the specification $\mathsf{Mult}(i_1, i_2, o, done)$:

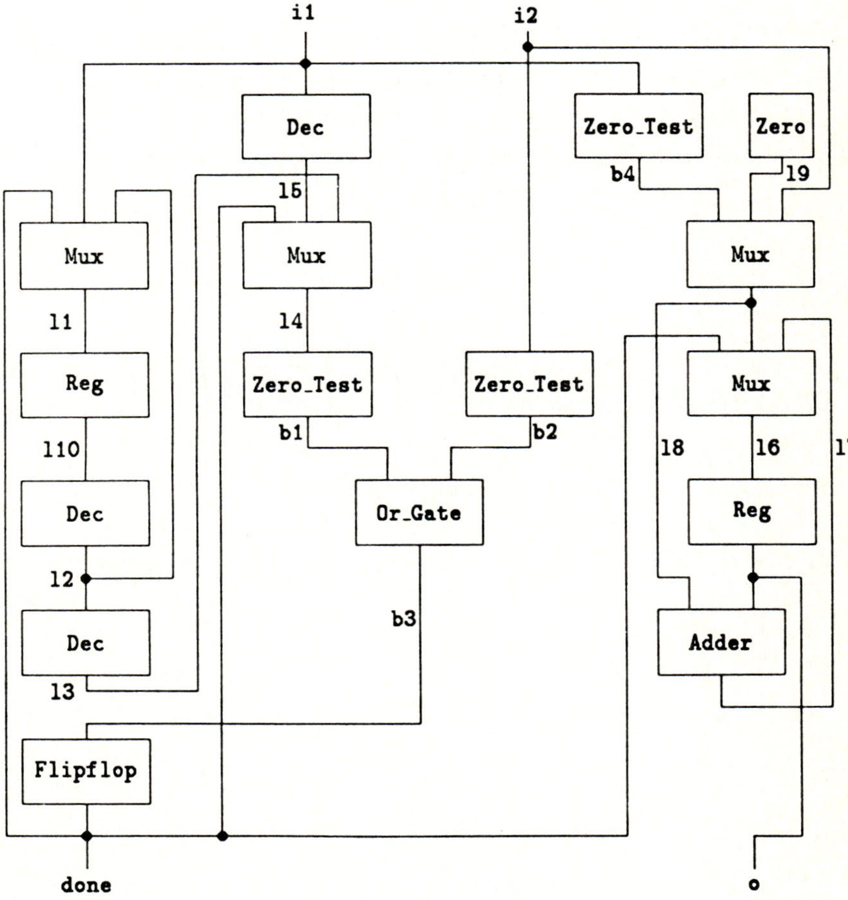

The components of this device are specified by:

$$\begin{array}{l}
\mathsf{Mux}(ctl, i_1, i_2, o) \ \equiv\ \forall t.\ o(t) = (ctl(t) \rightarrow i_1(t) \mid i_2(t)) \\
\mathsf{Reg}(i, o) \ \equiv\ \forall t.\ o(t{+}1) = i(t) \\
\mathsf{Flipflop}(i, o) \ \equiv\ \forall t.\ o(t{+}1) = i(t) \\
\mathsf{Dec}(i, o) \ \equiv\ \forall t.\ o(t) = i(t){-}1 \\
\mathsf{Add}(i_1, i_2, o) \ \equiv\ \forall t.\ o(t) = i_1(t) + i_2(t) \\
\mathsf{Zero_Test}(i, o) \ \equiv\ \forall t.\ o(t) = (i(t){=}0) \\
\mathsf{Or_Gate}(i_1, i_2, o) \ \equiv\ \forall t.\ o(t) = i_1(t) \vee i_2(t) \\
\mathsf{Zero}(o) \ \equiv\ \forall t.\ o(t) = 0
\end{array}$$

The behaviour corresponding to the diagram above is captured in logic as follows:

$$\begin{array}{l}
\mathsf{Mult_Imp}(i_1, i_2, o, done) \ \equiv \\
\exists b_1\ b_2\ b_3\ b_4\ l_1\ l_2\ l_3\ l_4\ l_5\ l_6\ l_7\ l_8\ l_9\ l_{10}. \\
\quad \mathsf{Mux}(done, l_8, l_7, l_6)\ \wedge\ \mathsf{Reg}(l_6, o)\ \wedge\ \mathsf{Add}(l_8, o, l_7)\ \wedge\ \mathsf{Dec}(i_1, l_5)\ \wedge \\
\quad \mathsf{Mux}(done, l_5, l_3, l_4)\ \wedge\ \mathsf{Mux}(done, i_1, l_2, l_1)\ \wedge\ \mathsf{Reg}(l_1, l_{10})\ \wedge \\
\quad \mathsf{Dec}(l_{10}, l_2)\ \wedge\ \mathsf{Dec}(l_2, l_3)\ \wedge\ \mathsf{Zero}(l_9)\ \wedge\ \mathsf{Mux}(b_4, l_9, i_2, l_8)\ \wedge \\
\quad \mathsf{Zero_Test}(i_1, b_4)\ \wedge\ \mathsf{Zero_Test}(l_4, b_1)\ \wedge\ \mathsf{Zero_Test}(i_2, b_2)\ \wedge \\
\quad \mathsf{Or_Gate}(b_1, b_2, b_3)\ \wedge\ \mathsf{Flipflop}(b_3, done)
\end{array}$$

9.3. Verification

The correctness of the multiplier implementation is established by proving that for all values of i_1, i_2, o and *done*:

$$\mathsf{Mult\_Imp}(i_1, i_2, o, done)\ \supset\ \mathsf{Mult}(i_1, i_2, o, done)$$

Expanding the definition of Mult and then slightly rearranging the result yields:

$$\begin{array}{l}
\forall t_1\ t_2.\ \mathsf{Mult_Imp}(i_1, i_2, o, done)\ \wedge \\
\qquad done(t_1)\ \wedge \\
\qquad \mathsf{Next}(t_1, t_2)(done)\ \wedge \\
\qquad \mathsf{Stable}(t_1, t_2)(i_1)\ \wedge \\
\qquad \mathsf{Stable}(t_1, t_2)(i_2)\ \wedge \\
\qquad \supset \\
\qquad (o(t_2) = i_1(t_1) \times i_2(t_1))
\end{array}$$

This can be proved by mathematical induction on $t_2 - t_1$. The proof is mostly routine, but there are a few slightly tricky bits. Some elementary results concerning $+$ and $\times$ are required, together with the following lemmas about time:

$$f(t+1) \supset \mathsf{Next}(t, t+1)(f)$$
$$\mathsf{Next}(t_1, t_2)(f) \wedge \neg f(t_1+1) \supset \mathsf{Next}(t_1+1, t_2)(f)$$
$$\mathsf{Next}(t_1, t_2)(f) \wedge \mathsf{Next}(t_1, t_3)(f) \supset (t_2 = t_3)$$
$$\mathsf{Next}(t, (t+1)+d)(f) \wedge \neg f(t+1) \supset \neg(d=0)$$
$$\mathsf{Stable}(t_1, t_2)(f) \supset \mathsf{Stable}(t_1+1, t_2)(f)$$
$$\mathsf{Stable}(t, (t+1)+d)(f) \wedge \neg(d=0) \supset (f(t) = f(t+1))$$

We do not give details of the proof here. It was not difficult to generate using the HOL system. Tom Melham has done a similar example: a device for computing the factorial function. In the process he discovered some general principles for reasoning about self-timed devices. Details will appear in a forthcoming paper.

10. An edge-triggered Dtype

The implementation of the multiplier described in the preceding section was described at the register-transfer level. This is an abstract level in which devices are viewed as sequential machines. At this level registers are modelled as unit-delay elements without explicit clock lines. To implement such a register using actual hardware, something like a Dtype device must be used:

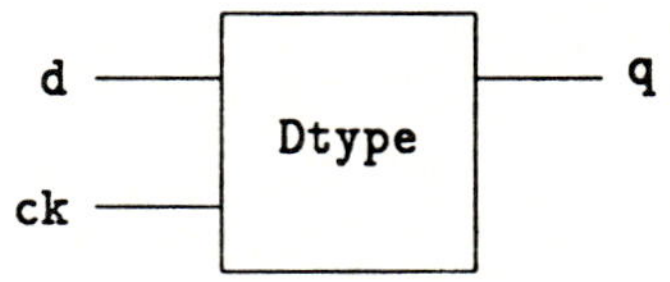

10.1. Specification

An informal behavioural specification of `Dtype` is:

If

- the clock `ck` has a rising edge at time t_1, and
- the next rising edge of `ck` is at t_2, and
- the value at `d` is stable for c_1 units of time before t_1 (c_1 is the setup time), and

- there are at least c_2 units of time between t_1 and t_2 (c_2 is the minimum clock period),

then

- the value at **q** will be stable from c_3 units of time after t_1 (c_3 is the start time) until c_4 units of time after t_2 (c_4 is the finish time), and
- the value at **q** between the start and finish times will equal the value held stable at **d** during the setup time.

To formalize this we need to define what a "rising edge" is. We will continue to use a discrete model of time, but the grain of time will be finer than before. A function from time to truth-values is defined to rise at time t if it is F at time $t-1$ and T at t. Formally:

$$\mathsf{Rise}(f)(t) \;\equiv\; (f(t-1) = \mathsf{F}) \;\wedge\; (f(t) = \mathsf{T})$$

If the function Rise is applied to a single argument f, then the resulting expression Rise(f) denotes a predicate that is true of t if and only if f rises at t. The specification of the Dtype below illustrates the use of this kind of 'partial application'.

The informal behavioural specification of a Dtype can now be formalized in logic by:

$$\begin{array}{l}\mathsf{Dtype}(c_1, c_2, c_3, c_4)(d, ck, q) \;\equiv \\ \forall t_1\; t_2.\; \mathsf{Rise}(ck)(t_1) \wedge \\ \qquad\quad \mathsf{Next}(t_1, t_2)(\mathsf{Rise}(ck)) \wedge \\ \qquad\quad \mathsf{Stable}(t_1 - c_1, t_1)(d) \\ \qquad\quad (t_2 - t_1 > c_2) \wedge \\ \qquad\quad \supset \\ \qquad\quad (\mathsf{Stable}(t_1 + c_3, t_2 + c_4)(q) \;\wedge\; (q(t_2) = d(t_1)))\end{array}$$

The parameters c_1, c_2, c_3 and c_4 are the timing constants of the Dtype; their value depends on how the device is fabricated. Note that Next(t_1, t_2)(Rise(ck)) is an expression formed by applying Next(t_1, t_2) to the predicate Rise(ck).

A Dtype becomes a unit-delay if we abstract signals to the sequence of values occurring at rising edges of the clock. The formal analysis of such abstractions is currently being studied by Tom Melham as part of his Ph.D. research.

10.2. Implementation

A common implementation of a Dtype uses NAND-gates:

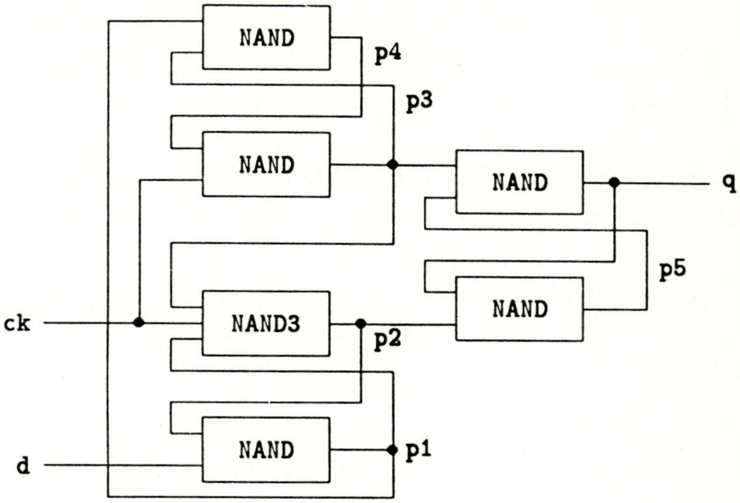

10.3. Verification

To show that this implementation works we must use a model in which the NAND-gates have delay, since it is the delay in feedback loops that provides memory. The simplest such model is one in which each gate has unit-delay:

$$\mathsf{NAND}(i_1, i_2, o) \equiv \forall t.\ o(t+1) = \neg(i_1(t) \wedge i_2(t))$$
$$\mathsf{NAND3}(i_1, i_2, i_3, o) \equiv \forall t.\ o(t+1) = \neg(i_1(t) \wedge i_2(t) \wedge i_3(t))$$

The Dtype implementation can be represented in logic by defining:

$$\begin{aligned}&\mathsf{Dtype\_Imp}(d, ck, q) \equiv \\ &\exists p_1\ p_2\ p_3\ p_4\ p_5. \\ &\quad \mathsf{NAND}(p_2, d, p_1) \wedge \mathsf{NAND3}(p_3, ck, p_1, p_2) \wedge \\ &\quad \mathsf{NAND}(p_4, ck, p_3) \wedge \mathsf{NAND}(p_1, p_3, p_4) \wedge \\ &\quad \mathsf{NAND}(p_3, p_5, q) \wedge \mathsf{NAND}(q, p_2, p_5)\end{aligned}$$

It can then be proved that:

$$\mathsf{Dtype\_Imp}(d, ck, q) \supset \mathsf{Dtype}(2,3,4,1)(d, ck, q)$$

This shows that if each NAND-gate has unit-delay then the Dtype has a setup time of 2, a minimum clock period of 3, a **start** time of 4 and a finish time of 1. The formal proof of this is fairly complicated. It has been done by hand by John Herbert.

11. Conclusions

The examples presented here demonstrate that higher-order logic is a formalism in which a wide variety of behaviour and structure can be specified. With the aid of suitable functions (*e.g.* Iterate), specifications can be made to look like conventional hardware descriptions whilst retaining logical purity (and hence formal tractability).

Hardware verification requires various kinds of reasoning.

- The adder example shows the need for mathematical induction (both to deal with iterated structures and for proving arithmetic lemmas).
- The multiplier example shows the need for reasoning about temporal concepts (Next, Stable *etc.*).
- The Dtype and unit-delay show the need for reasoning about abstractions between different time scales.

All these kinds of reasoning can be done using the standard inference rules of logic.

12. Acknowledgements

The use of higher-order logic for hardware specification and verification has been pioneered by Keith Hanna [7]. Many of the techniques presented here have been adapted from Ben Moszkowski's work on applying temporal logic to hardware description [6]. In particular, I learnt from him the representation of circuits as predicates described in Section 4; this is a big improvement over the *ad hoc* representation I had previously been using [3]. I have had numerous useful discussions with the users of the HOL system. These include Albert Camilleri, Nives Chaplin, Inder Dhingra, John Herbert, Tom Melham and Edmund Ronald, all from Cambridge, and Jeff Joyce from the University of Calgary. Don Gaubatz, Andy Hopper and Peter Robinson patiently explained to me various electrical phenomena that arise in MOS circuits. Avra Cohn, Inder Dhingra and John Herbert pointed out errors in a first draft of this paper.

13. References

[1] Church, A., *A Formulation of the Simple Theory of Types*, Journal of Symbolic Logic 5, 1940.

[2] Clocksin, W. F., *Logic Programming and the Specification of Circuits*, Computer Laboratory Technical Report No. 72, 1985.

[3] Gordon, M. J. C., *LCF_LSM*, University of Cambridge Computer Laboratory Technical Report No. 41, 1983.

[4] Gordon, M. J. C., *HOL: A Machine Oriented Formulation of Higher-Order Logic*, University of Cambridge Computer Laboratory Technical Report No. 68, 1985.

[5] Gordon, M. J. C., *HOL: A Proof Generating System for Higher-Order Logic*, Forthcoming technical report.

[6] Halpern, J., Manna, Z. and Moszkowski, B., *A Hardware Semantics based on Temporal Intervals*, in: *Proceedings of the 10-th International Colloquium on Automata, Languages and Programming*, Barcelona, Spain, 1983.

[7] Hanna, F. K. and Daeche, N., *Specification and Verification using Higher-Order Logic*, in: *Proceedings of the 7-th International Conference on Computer Hardware Design Languages*, Tokyo, 1985.

[8] Hatcher, W., *The Logical Foundations of Mathematics*, Pergamon Press, 1982.

[9] Hoare, C. A. R., *A Calculus of Total Correctness for Communicating Processes*, Science of Computer Programming, Vol. 1, No. 1, 1981.

[10] *Designing with Gate Arrays.* Lattice Logic Limited, 9 Wemyss Place, Edinburgh EH3 6DH, 1982.

[11] Milner, R., *A Calculus of Communicating Systems*, Lecture Notes in Computer Science No. 92. Springer-Verlag, 1980.

Formal Aspects of VLSI Design
G.J. Milne and P.A. Subrahmanyam (editors)

Specification and Verification using Higher-Order Logic: A Case Study

F K Hanna and N Daeche

University of Kent
Canterbury, Kent, U.K.

Higher-order logic is an ideal computer hardware description language: it allows the behaviour of digital devices to be rigorously specified, and it allows the correctness of digital system designs to be formally verified. The VERITAS approach to specification and verification is based on the use of higher-order logic. As a case study of this approach, a behavioural specification for an edge-triggered D-type flipflop is formulated, and a description is given of the stages involved in computationally verifying the correctness of a commonly used implementation of this kind of flipflop.

1. Introduction

Contemporary computer hardware design languages (CHDL's) have been designed to achieve many aims, including the ability to:

- describe the *structure* of digital systems (possibly hierarchically),
- describe the *behaviour* of digital systems (possibly at different levels of abstraction),
- *simulate* the behaviour of assemblages of digital components,
- *transform* structural or behavioural descriptions into other forms (for example, into wiring lists, or into waveform diagrams, etc),
- generate *fault diagnostic* information, etc.

One aim, however, which by and large they have *not* set out to achieve is that of allowing *reasoning* about the properties of structures, the properties of behaviours, or of the interrelationship between structures and behaviours. Alternatively stated, little emphasis has been given to the definition of the *semantics* of CHDL's.

In this respect, CHDL's are similar to many existing programming languages (for example, FORTRAN, LISP, ADA, etc). These languages, although they are seen as being well suited for expressing complex sequences of computation, have semantics which (to the extent that they have been defined at all !) are of immense intricacy. For example, a recent attempt to define the semantics of ADA ran to some several hundred pages...

On the other hand, some recent programming languages (OCCAM being a prime example) do have a well defined, and relatively simple, semantics. This comes about not by accident, but because the design of such languages *started* with semantic considerations, with other aspects of the languages (eg, syntax, implementation, etc) being attended to only *after* the semantics had been rigorously defined. The end result is that it is relatively simple to reason (either intuitively, or formally) about the properties (correctness, termination, equivalence, etc) of programs written in such languages.

We believe that a similar approach can usefully be adopted towards the design of CHDL's. The clarity, simplicity and rigour of definition of the *semantics* of a CHDL (rather than, for example, its syntax, or its built-in timing primitives, etc) should be *the prime consideration* influencing its design.

Once this point of view has been adopted, the question naturally arises as to what notation or formalism is best suited for expressing the semantics of a CHDL. It is not difficult to list the relevant criteria:

- Ideally, the formalism should *already exist*. This means that it will have been well researched, (independently of our intended use for it) and there will be an existing community of 'experts' from whom help may be sought in resolving any tricky technical points.
- The formalism should be both *powerful* and *concise*. In fact, if we intend to use it to reason about arbitrary properties of complex hardware designs, it should be powerful enough to encompass a *large part of mathematics itself.*
- At the same time, the formalism should be not too far removed from the informal modes of expression and intuitive reasoning that digital engineers use most of the time.
- It should allow *partial* descriptions to be expressed and reasoned about. The real world is both indeterminate and of unbounded complexity; our knowledge of it is therefore *inherently* partial. Thus, any *total* description of it must necessarily be an incorrect one.
- Finally, the formalism should be a strictly *formal* one. This implies that reasoning in the formalism can be carried out by *formal inferencing* (or, alternatively expressed, by symbol manipulation). In turn, this implies that the logical soundness of any reasoning carried out in the formalism can be *computationally checked.*

Taken together, this list of criteria is almost prescriptive in suggesting *typed, higher-order predicate logic* [11] as the formalism of choice. As a brief illustration of why *higher*-order logic, rather than just *first*-order logic is appropriate, consider the following sequence of concepts:

- The *voltage* at a port. We can consider a voltage to be a *first-order* entity.
- The *waveform* at a port. A waveform is a function from time to voltage, and hence it is a *second-order* entity.
- The *behavioural specification* of a device. As we shall discuss in §2.3 below, a behavioural specification is a predicate on a tuple of waveforms. It is, therefore a *third-order* entity.
- The relation '*better than*', meaning that one behavioural specification is stronger (ie, more

restrictive) than another. This is a predicate on pairs of behavioural specifications, and hence is a *fourth-order* entity.

- The property '*is a transitive relation*' (a property that we might wish to assert the above relation '*better than*' possesses). This property is a predicate on the above fourth-order relation, and hence it itself is a *fifth-order* entity.

Thus, (and this is not a particularly 'contrived' example) we find that a fairly intuitive property — the transitivity of the above relation — is a fifth-order one. By using higher-order logic we gain the ability to freely introduce and manipulate higher-order entities. In turn, this leads to an easy and natural correspondence between intuitive, everyday concepts and their precise, formal expression within the CHDL.

The reason why a 'typed' species of logic is used is twofold. Firstly, the (informal) notations of science and engineering tend to be typed (eg, *let w be a waveform, t be an instant of time,* etc) and thus typed logic provides a more concise and natural means of expression than does untyped logic. Secondly, like the untyped λ-calculus, 'ordinary' higher-order logic suffers from being inconsistent; the introduction of types is one way of overcoming this defect.

Relation to other formalisms Another feature we might mention in favour of the use of typed higher-order logic is that it includes, as proper subsets, two other formalisms which are also currently being explored as a basis for reasoning about the properties of digital systems. These are:

- The 'functional' subset (essentially Church's typed λ-calculus). Terms in this subset can be 'evaluated', and thus behavioural specifications restricted to this subset can be 'animated'.
- The 'first-order, Horn clause' subset (essentially PROLOG). The truth values of (some) terms expressed in this subset can be mechanically determined, and thus questions posed about properties of behavioural specifications restricted to this subset can be automatically answered.

1.1 Background

We have given the name VERITAS to the general approach to specification and verification of digital systems based on the use of predicate logic. We also use this name variously to describe the overall methodology, the project under which this approach has been developed, and the particular species of logic used. Work on the VERITAS project started in 1981, with the publication [2] of a set of theory presentations characterising, within *first*-order logic, the structural and behavioural properties of discrete logic devices, and the first computationally checked proofs were obtained in 1982. The use of *higher*-order logic for this purpose was first proposed in [3].

Relation to other work The use of logic to describe properties of the physical world and to reason about their interrelation finds its origins with Euclid's informal axiomatisation of geometry in ancient times, and with Hilbert's more formal axiomatisation of the same at the beginning of this century. Since then, propositional logic (Boolean algebra) has underpinned much of the development of digital logic and computers, and, over the last couple of decades, both first-order logic and the λ-calculus have found extensive use in software verification.

Within the field of hardware verification, approaches relating to the one described here have been reported by several authors. Eveking [8] has advocated the use of first-order logic and has developed an approach based on the use of predicate transformers, and has discussed the way that descriptions at different levels of abstraction may be related. The use of higher-order logic (with an emphasis towards specification and verification at higher levels of abstraction than those discussed here) is described [7] by Gordon, and an associated case study by Herbert is presented in [9]. Finally, we mention that the use of predicate logic to reason about circuits at the analogue level is described [6] by Shostak.

1.2 Contents

The main aim of this paper is to present a case study, of moderate complexity, demonstrating the VERITAS approach to the behavioural specification of a digital device, and to the formal verification of a proposed implementation of it. The remainder of this paper is organised as follows:

- We begin with a discussion of our approach to *partially describing* waveforms, and introduce the concept of a '*wavespec*' — that is, a waveform specification.
- We illustrate (as a simple example) how, using higher-order logic, the behaviour deemed to characterise a NAND gate can be specified.
- As a more complex example, we specify the behaviour of an edge-triggered, D-type flipflop (we abbreviate the name of this device to 'D-flipflop').
- We describe a proposed implementation, in terms of NAND gates, of a D-flipflop, and *attempt* to describe, at an intuitive level, how it works. This turns out not to be easy ...
- We introduce the notions of *formal verification*, and of *computational theorem proving*.
- As a case study in formal verification, we describe the various stages involved in computationally constructing a formal proof asserting the correctness of the above proposed implementation of the D-flipflop.
- Finally, we indicate the practical significance of the above result, and summarise our main conclusions.

Although this paper is largely self-contained, it does not attempt to cover all aspects of the subject. A different perspective on the use of higher-order logic for specification and verification will be found in [4], with an emphasis on the formal specification of structures (ie, circuit diagrams). In addition, in [5] can be found a full and detailed account of the formal logic used here and a description of the VERITAS Theorem Prover, together with a tutorial exposition on the latter.

1.3 Notation

The mathematical notation used in this account is, with one or two minor exceptions, one of the notations used by the computational implementation of the system (as described later). There are only three features that need comment:

- The *application* of a function f to a set of arguments $x_1, x_2, \ldots, x_n$ is denoted (in the 'applicative' style) by $(f \quad x_1\ x_2 \ldots x_n)$ rather than by $f(x_1, x_2, \ldots, x_n)$.
- It is often convenient to represent multi-argument functions in higher-order (or 'curried') form; for example $((F\ x_1)\ x_2)$ instead of $(f\ x_1\ x_2)$.
- The relative indentation of multi-line expressions is used, along with parentheses, to indicate grouping.

2. General Approach

The methodological aspects of the VERITAS approach are a little more fully described in [4]; here we have space to present only a very brief and superficial summary. Overall, the approach involves defining a *theory* which characterises the universe of discourse of digital engineering. A theory is defined by a *signature*; this is simply a list of the *symbols* that can occur in terms and the *axioms* that can be used in constructing *derivations* (or 'proofs') of theorems within the theory. The intention is that:

- Each symbol should correspond to some entity in the universe of discourse.
- Each axiom should describe some relation between these entities that is known (or asserted) to be true.

In practice, it is convenient to define a signature in a structured way, as a series of *extensions* of a primitive signature. This therefore gives rise to a corresponding nested series of theories. In the following subsections, we summarise the principal features of each of these theories and indicate their intent. The most primitive theory in the sequence is that of pure higher-order logic. The symbols introduced include the type *bool* (comprising the two propositional truth values *true* and *false*) together with the logical connectives ($\wedge$, $\vee$, $\neg$, etc) and equality (ie, '='). The axioms introduced define certain fundamental properties of higher-order logic (for instance, reflexivity of equality, extensionality, etc).

2.1 Theory of Numbers and of Time

The first important extension of the signature of pure higher-order logic defines the theory of Natural Numbers, introducing the type *nat* (of natural numbers), various arithmetic operators (*succ*, $+$, $-$, etc), and a set of axioms that characterise number theory. As an example, the following axiom describes induction:

$$\begin{aligned}&\vdash \forall P : nat \to bool.\\ &\qquad (P \quad 0)\\ &\quad \to\\ &\qquad \forall n : nat.\ (P \quad n) \to (P \quad (succ \quad n))\\ &\quad \to\\ &\qquad \forall n : nat.\ (P \quad n)\end{aligned}$$

(Note that, since this axiom involves quantification over the set of all properties P, it is an intrinsically higher-order one.)

One feature of this theory to which we need to draw attention is that the operator '$-$' denotes *natural subtraction* (that is, $a - b = 0$ if $b \geq a$). This operator shares only some of the properties of 'ordinary' subtraction (as commonly defined on the integers). As a simple example, consider the relation $(a + b) - c = (a - c) + b$ Whilst this relation is true if $b = 0$ or if $a \geq c$, it is *not* true in general.

The next extension of the signature defines a theory of time. Specifically, it introduces three data types

time — describing *instants* of time
dur — describing non-negative *durations* of time
intvl — describing *intervals* of time

and a collection of functions between them. In particular, the function

$$I : \ time \times dur \to intvl$$

is used for constructing intervals. The interval $(I \quad t \quad d)$ is defined as the semi-open interval starting at t, and of duration d.

2.2 Theory of Waveforms

The behaviour of a digital system is defined in terms of the waveforms present at its ports. Although these are often idealised as being perfect, two-valued 'binary' waveforms, in reality they are *analogue* waveforms that are only in a 'binary' state during certain intervals of time. The general approach followed by the VERITAS axiomatisation is not to pretend that reality is other than it is, but simply to axiomatise *only* those properties of it that we actually require.

In the theory of Waveforms, defined by the next extension of the signature, two types are introduced; these are wf the set of *analogue* waveforms, and C, a type containing the two elements c_f and c_t, representing the two 'digital' constants. (Note: these digital constants, (pseudonyms *0*, *1*, or *low*, *high*, etc) should not be confused in any way with the propositional truth values *true*, *false* of type *bool* introduced above).

Wavespecs One of the foundations upon which the entire VERITAS approach to specifying behaviours rests is in the use of a particular predicate

$$const : (intvl \times C) \to (wf \to bool)$$

This predicate (the fact that it is 'curried', or higher-order, is unimportant) is defined by the term

$$((const \quad i \quad c) \quad w)$$

meaning that throughout an interval of time i, the analogue waveform w assumes the digital value c. This is sketched in Fig. 1a. Note the *partial* nature of *const*, and in particular, the fact that it makes no assertion whatsoever concerning the behaviour of w outside the interval i. It is convenient to introduce a graphical notation, as shown in Fig. 1b, for expressing assertions of this general form.

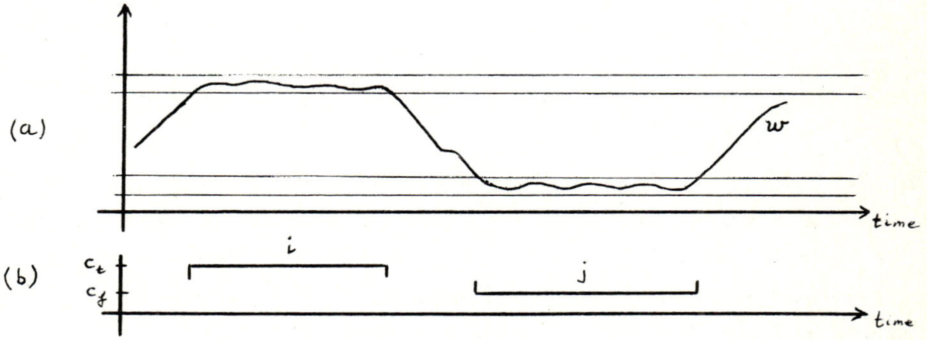

Fig 1.

(a) Both the assertion $((const \quad i \quad c_t) \quad w)$ and the assertion $((const \quad j \quad c_f) \quad w)$ are true.

(b) This shows the convention adopted for depicting these assertions graphically.

We call a predicate defined on a waveform (or, more generally, on a collection of waveforms) a *wavespec* (for 'waveform specification'). For example, the term $(const \quad i \quad c_t)$ is a wavespec. As we shall see, wavespecs are of fundamental importance in the VERITAS approach.

2.3 Theory of Gate Behaviours

The theory of Gate Behaviours (defined by a further extension of the above signature) contains the definitions for the *behavioural* (as distinct from the *structural*) characteristics implied by common usage of the terms 'AND gate', 'OR gate', etc. Insofar as these are pure *definitions*, the reader is free to decide the extent to which they succeed in capturing what is generally meant by these terms.

We will illustrate the general approach adopted for characterising device behaviours by taking the behaviour of a 2-input NAND gate as an example: other gate behaviours follow a similar pattern, as do other device behaviours (e.g. RAM's, ALU's, etc).

We begin by noting that, as a NAND gate has three ports, the general form of the definition will be a predicate on three waveforms (ie, a wavespec on a 3-tuple of waveforms). Because NAND gates have differing 'speeds', the definition must also incorporate some *timing parameters*. We choose to use a tuple of four durations to characterise the speed; this allows a much 'tighter' definition than is normally used for specifying the characteristics of gates. As an aside, we note that there is, of course, no reason why a 'looser' definition (ie, one with fewer parameters) should not be used if so desired, but then, because less would be known about the gate's behaviour, it would not be possible to infer so much about the properties of systems employing such gates.

The symbol we use for characterising the behaviour of a NAND gate is

$$NAND\_BEHAV : dur^4 \rightarrow (wf^3 \rightarrow bool)$$

As before, it turns out that it is convenient to express it as a *higher-order* function. The meaning we wish the term

$$((NAND\_BEHAV \quad d_1 \quad d_2 \quad d_3 \quad d_4) \quad w_{in_1} \quad w_{in_2} \quad w_{out})$$

to have is (approximately):

"The three waveforms w_{in_1}, w_{in_2} and w_{out} exhibit the behavioural characteristics of a NAND gate, with propagation and hold times (for *low* inputs) of d_1 and d_2 respectively, and propagation and hold times (for *high* inputs) of d_3 and d_4 respectively."

Let us now propose a definition for this predicate. We will first illustrate its intent in terms of waveform sketches, and then evolve the corresponding formal definition.

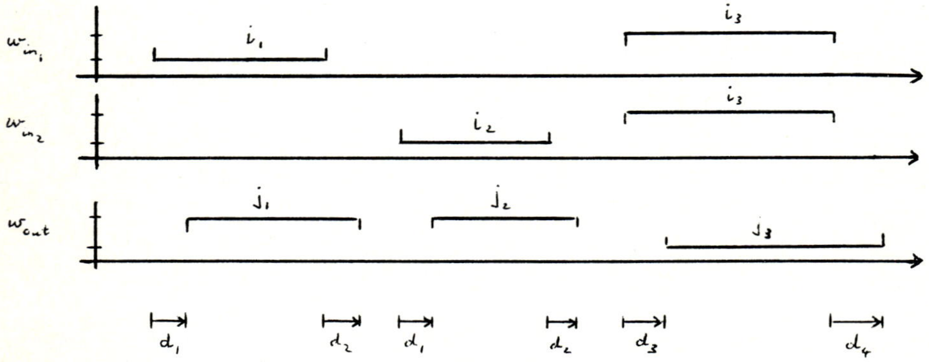

Fig 2. Illustration of the predicate $((NAND\_BEHAV \quad d_1 \; d_2 \; d_3 \; d_4) \quad w_{in_1} \; w_{in_2} \; w_{out})$

Let w_{in_1} and w_{in_2} be the waveforms at the input ports of a NAND gate, and assume (see Fig 2) that these two waveforms satisfy the various constraints shown in the diagram (ie, the waveform w_{in_1} is in the c_f (or '*low*') state throughout the interval i_1, etc). Consider what assertions may be made about the waveform w_{out} at the output port of the gate. We would argue that (to capture the essence of 'NAND gate behaviour') this waveform must satisfy the constraints shown (ie, it must be in the c_t state throughout an interval j_1, which starts d_1 after i_1 starts and finishes d_2 after i_1 finishes, etc).

There are three interesting points about this definition. First, it is a *partial* definition of the behaviour: it places bounds on what is required, but it does not artificially overconstrain it. For example, it does not imply idealised rise and fall transients. Secondly, it is *non-strict* for 'low' inputs: for example, during interval i_1 it does not impose *any* requirements on the behaviour of waveform w_{in_2}. Thirdly, it does not state what the *actual* rise and fall times of the analogue waveforms must be: it simply specifies bounds on them.

Now let us make this definition precise. First, it is useful to introduce, as a subsidiary definition, a function $\Delta : dur^2 \rightarrow (intvl \rightarrow intvl)$ that 'propagates' intervals. The interval

$$((\Delta \quad d_1 \quad d_2) \quad i_1)$$

is *defined* to be the interval which starts d_1 later that i_1 starts and finishes d_2 later than i_1 finishes (it is, in fact, the interval j_1 of Fig(2)). Using this function, we can define the 'low-mode' aspects of the $NAND\_BEHAV$ predicate, as follows:

$$\begin{array}{l} \forall i : intvl. \\ \quad ((const \quad i \quad c_f) \quad w_{in_1}) \vee ((const \quad i \quad c_f) \quad w_{in_2}) \\ \quad \rightarrow \\ \quad ((const \quad ((\Delta \quad d_1 \quad d_2) \quad i) \quad c_t) \quad w_{out}) \end{array}$$

or, in narrative form:

"**For any** interval i, **if** waveform w_{in_1} is in state c_f throughout i, **or** waveform w_{in_2} is in state c_f throughout i, **then** waveform w_{out} will be in state c_t throughout the interval starting d_1 after i and finishing d_2 after i."

The corresponding 'high-mode' aspects is identical, save only that the roles of c_f and c_t are interchanged, that d_3, d_4 appear in place of d_1, d_2, and that $\vee$ (disjunction) is replaced by $\wedge$ (conjunction).

We can now present the actual definition of the behavioural predicate $NAND\_BEHAV$; it is

$$\begin{array}{l} ((NAND\_BEHAV \quad d_1 \quad d_2 \quad d_3 \quad d_4) \quad w_{in_1} \quad w_{in_2} \quad w_{out}) \\ =_{def} \\ \forall i : intvl. \\ \quad ((const \quad i \quad c_f) \quad w_{in_1}) \vee ((const \quad i \quad c_f) \quad w_{in_2}) \\ \quad \rightarrow \\ \quad ((const \quad ((\Delta \quad d_1 \quad d_2) \quad i) \quad c_t) \quad w_{out}) \\ \wedge \\ \quad ((const \quad i \quad c_t) \quad w_{in_1}) \wedge ((const \quad i \quad c_t) \quad w_{in_2}) \\ \quad \rightarrow \\ \quad ((const \quad ((\Delta \quad d_3 \quad d_4) \quad i) \quad c_f) \quad w_{out}) \end{array}$$

In narrative form, this reads (approximately) as:

"The predicate $NAND\_BEHAV$ (that is deemed to characterise the behaviour of a NAND gate) applied to a 4-tuple of durations d_1, d_2, d_3, d_4 (representing the timing parameters of the gate), in turn applied to a 3-tuple of waveforms $w_{in_1}, w_{in_2}, w_{out}$ (representing the waveforms at the ports of the gate) **is defined as meaning** that **for any** interval of time, i, the two following assertions hold

(1) **if** input waveform w_{in_1} is in state c_f throughout i, **or** input waveform w_{in_2} is in state c_f throughout i, **then** output waveform w_{out} will be in state c_t throughout the interval starting d_1 after i starts, and finishing d_2 after i finishes, **and**

(2) **if** input waveform w_{in_1} is in state c_t throughout i, **and** input waveform w_{in_2} is in state c_t throughout i, **then** output waveform w_{out} will be in state c_f throughout the interval starting d_3 after i starts, and finishing d_4 after i finishes.

There are a number of points to make in connection with this definition.

- It is quite a complicated definition (although, with familiarity, it is readily comprehensible). However, *we would maintain that this complexity is intrinsic to the digital designers notion of the behaviour of* a NAND *gate*.
- One point we have deliberately glossed over is the *partial* nature of Δ, the 'propagate' function. In particular, we would *not* wish to be able to apply Δ to a *null* interval and have it yield a *non-null* interval. For example, we would not wish the interval defined by

$$((\Delta \quad 2 \quad 5) \quad (I \quad t \quad 0))$$

to be the interval $(I \quad t+2 \quad 3)$. In the present context, such behaviour (corresponding to 'pulse-stretching' a null pulse into a non-null one) would be distinctly non-physical! The axiomatic definition of Δ has been carefully contrived to avoid such a meaning. (In this respect, it differs from the definition of Δ as used in [4], in which (because all intervals were defined as being *closed* rather than semi-open, as here) it was necessary to 'guard' explicitly against such a possibility at every usage of Δ, a feature which gave behavioural definitions a somewhat clumsy appearance.)

3. Specification of a D-flipflop

In this section we will first describe, in informal terms, how (we believe) an 'edge-triggered, D-type flipflop' (or, in short, a D-flipflop) is meant to behave. We will then make this description more precise by recasting it as a collection of wavespec diagrams. Finally, we will formulate it as a wavespec *definition*.

3.1 Informal description

A D-flipflop (see Fig. 3a) has three ports: a 'strobe' (or clock) port (S), a 'data-in' port (D) and an 'output' port (Q). It is typically described in textbooks and manufacturer's specification sheets by a set of idealised waveform diagrams such as those shown in Fig. 3b. Data (on the D port) is strobed on the rising edge of a strobe pulse and transferred shortly afterwards to the Q port, where it remains until the next strobe pulse. A manufacturer's data sheet (eg [10]) will specify certain conditions that have to be satisfied, and certain parameters that are guaranteed.

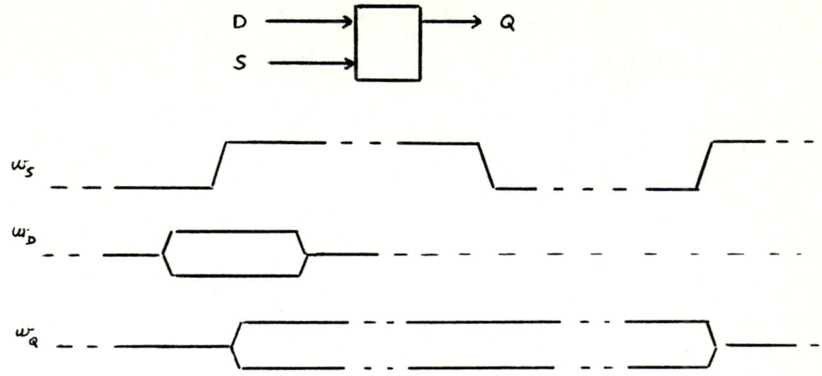

Fig 3.

(a) Graphical representation of a D-flipflop. S is the '*strobe*' (or *clock*) port, D is the '*data*' port, and Q is the '*output*' port.

(b) A typical 'idealised' behavioural description for a D-flipflop.

In the first category will (typically) be found maximum strobe rates and minimum data setup and hold times, and in the second category will (typically) be found maximum propagation delays.

3.2 Wavespec diagram description

Now consider a rather more careful description of the same device. As before (with the NAND gate), we emphasise that we are now attempting to formalise a *definition* of an already existing concept: the reader must decide an the extent to which he or she thinks we succeed in capturing the intent of the existing usage of this term. In passing, we note that we could (very easily!) devise a more complicated (or 'tighter fitting') definition than the one we are about to propose (for instance, we *could* decide to associate *different* timing parameters with different polarities of signals, etc). In the interests of keeping the exposition clear, however, we have deliberately chosen the most simple (yet realistic) definition possible. The definition we propose is of the form

$$\forall c : C.$$
$$(z_S \quad w_S) \wedge ((z_D \quad c) \quad w_D) \rightarrow ((z_Q \quad c) \quad w_Q)$$

Before describing the significance of this expression as a whole, first consider its component wavespecs, z_S, z_D etc. These are shown diagrammatically in Fig. 4

- z_S is a wavespec on the strobe waveform. It asserts that it must be *low* throughout the interval of duration a_1, and be *high* throughout an interval of duration of $a_3 + d$ (where d is a duration of arbitrary extent), and be *low* throughout an interval of duration e (where e is of arbitrary extent). Elsewhere the strobe waveform is unrestricted. Informally, we would say that this assertion describes a 'rising edge' on the strobe waveform, (with the

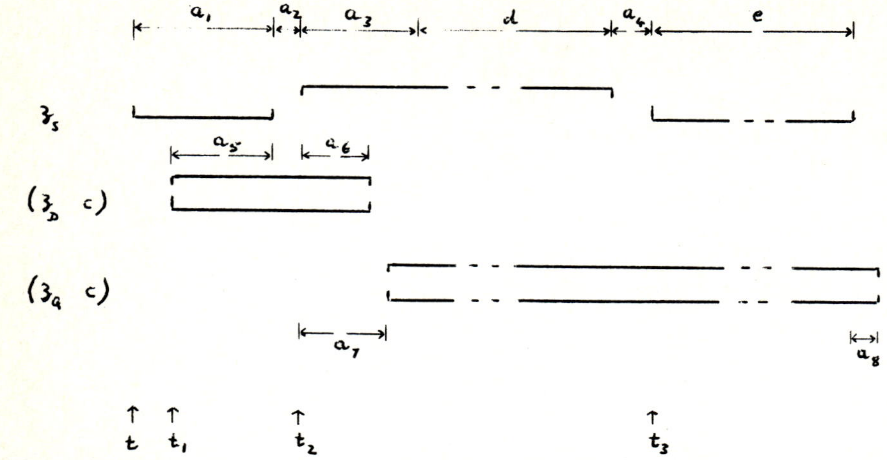

Fig 4. Wavespec definition of a D-flipflop. The timing parameters $\vec{a}$ correspond (loosely) to the following:

a_1	min strobe *low* time
a_2	max strobe rise time
a_3	min strobe *high* time
a_4	max strobe fall time
a_5	data setup time
a_6	data hold time
a_7	max output propagation delay
a_8	min output hold time

duration a_2 describing the maximum allowable rise time) followed by (if e is nonzero) a 'falling edge' (with maximum fall time a_4). Note carefully that the specification says nothing about the detailed 'shape' of either transition.

- $(z_D \quad c)$ is a wavespec on the data waveform. It asserts that the data waveform must be constant, at value c (where c is either c_t or c_f) throughout an interval which overlaps the 'rising edge' of the strobe waveform. The two duration a_5 and a_6 correspond to the required 'setup' and 'hold' times for the D-flipflop.
- $(z_Q \quad c)$ is a wavespec on the output waveform. It asserts that the output waveform w_Q is constant, at value c, throughout the interval shown. The two durations a_7 and a_8 describe the (maximum) propagation time and (minimum) hold time of the D-flipflop.

With these definitions, we can now appreciate the general form of the flipflop specification

$$\forall c : C.$$
$$(z_S \quad w_S) \wedge ((z_D \quad c) \quad w_D) \rightarrow ((z_Q \quad c) \quad w_Q)$$

In words "**For all** values of c (namely, either c_t or c_f), **if** w_S, the strobe waveform, satisfies wavespec z_S, **and** w_D, the data waveform, satisfies wavespec $(z_D \ \ c)$, **then** w_Q, the output waveform, will satisfy wavespec $(z_Q \ \ c)$".

In passing, we note that, as a special case, the duration e may be zero: if this is so, then the definition will not require there to be a well-defined falling edge on the strobe waveform.

3.3 Formal definition

Now that we have described (informally) our understanding of the term "*The behaviour of a D-flipflop*", let us propose a formal definition for it. To this end, we introduce a new symbol

$$DFF\_BEHAV : dur^8 \rightarrow (wf^3 \rightarrow bool)$$

with the intention that

$$((DFF\_BEHAV \ \ \vec{a}) \ \ w_S \ \ w_D \ \ w_Q)$$

should express the assertion: "*The three waveforms w_S (the strobe waveform), w_D (the data waveform) and w_Q (the output waveform) exhibit the behavioural characteristics of a D-flipflop with timing parameters $\vec{a}$* ". Formally, we define this symbol by the axiom

$$\begin{array}{l}
\vdash \ ((DFF_BEHAV \ \ \vec{a}\,) \ \ w_S \ \ w_D \ \ w_Q) \\
\quad =_{def} \\
\quad\ \forall c : C. \\
\qquad \forall t : time, \ d, e : dur. \\
\qquad\quad (z_S \ \ w_S) \wedge ((z_D \ \ c) \ \ w_D) \rightarrow ((z_Q \ \ c) \ \ w_Q) \\
\qquad WHERE \\
\qquad\quad (z_S \ \ w) \ = \\
\qquad\qquad ((const \ \ (I \ \ t \ \ a_1) \ \ c_f) \ \ w) \wedge \\
\qquad\qquad ((const \ \ (I \ \ t_2 \ \ a_3 + d) \ \ c_t) \ \ w) \wedge \\
\qquad\qquad ((const \ \ (I \ \ t_3 \ \ e) \ \ c_f) \ \ w) \\
\qquad\quad ((z_D \ \ c) \ \ w) \ = \\
\qquad\qquad ((const \ \ (I \ \ t_1 \ \ a_5 + a_2 + a_6) \ \ c) \ \ w) \\
\qquad\quad ((z_Q \ \ c) \ \ w) \ = \\
\qquad\qquad ((const \ \ ((\Delta \ \ a_7 \ \ a_8) \ \ i) \ \ c) \ \ w) \\
\qquad WHERE \\
\qquad\quad i = (I \ \ t_2 \ \ a_3 + d + a_4 + e) \\
\qquad\quad t_3 = t_2 + a_3 + d + a_4 \\
\qquad WHERE \\
\qquad\quad t_1 = t + a_1 - a_5 \\
\qquad\quad t_2 = t + a_1 + a_2
\end{array}$$

(**Note**: The locally defined symbols t_1, t_2, t_3 and i were indicated in Fig. 4)

As with the definition of $NAND\_BEHAV$, we note that this definition is moderately complex. It is, however, just about the *simplest possible* definition that captures the digital designer's notion of what constitutes the behaviour of a D-flipflop. Further, we note that whilst it fully specifies it, it does *not* overspecify it. This is an important point: overspecification will in general limit the range of possible implementations and therefore unnecessarily increase implementation costs.

Standard Definitions As a final point, we note that, in formulating this definition, several choices had to exercised, as anyone starting off to frame such a definition *ab initio* will very soon discover! Thus, whilst we believe that the proposed definition is a reasonable one, it is not unique: many subtle variations are possible. One might envisage, one day, a forward looking standards organisation (BSI, IEEE, ISO, etc) taking on the task of formulating a set of 'standard' definitions for the elements of digital systems, much as Whitworth did for screw threads, or Bourbaki did for mathematics. In the shorter term, one might envisage individual companies each evolving their own 'house style', a shared set of definitions that would be available to anyone in the company who was involved in specification, implementation or verification. Such a 'database' (which would, in time, include derivations of useful lemmas as well) would be a valuable resource.

4. Implementation

Now that we have defined the behaviour that (at least in our eyes) a D-type flipflop is *meant* to exhibit, we will consider a possible implementation of one (there is, of course, an infinity of means by which such behaviours *could* be realised). The implementation we choose to examine is shown in Fig. 5. We came across it in the reference manual for a ULA manufacturer's cell library. Prior to that the same circuit may be found in [10] as a 'functional block diagram' for a SN7474 integrated circuit.

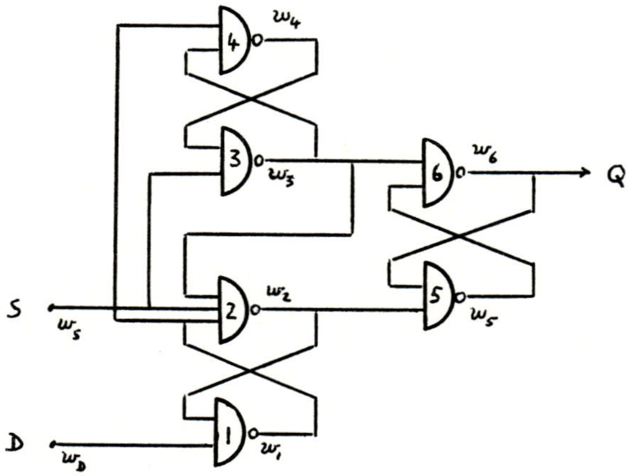

Fig 5. The circuit of a widely used implementation of a D-flipflop.

4.1 How it appears to work

Before undertaking (in §5) a formal proof that this circuit actually exhibits the behaviour of a D-flipflop, it naturally makes sense to try to gain some *intuitive* understanding of how the circuit operates. We say 'try' because, perhaps surprisingly in view of the apparent simplicity of the circuit, this is very difficult! It turns out, on analysis, that the *modus operandi* of this circuit is far from simple: in fact, it is unusually complex, and (so the authors found) difficult to understand intuitively*.

It turns out that the circuit operation can be understood most easily by considering the circuit as two subcircuits, of which the second can be treated as an ordinary set-reset flipflop (see Fig. 6a). This then allows the operation of the circuit to be appreciated in higher-level terms. Lest this manœuvre seem so commonplace as barely to warrant comment, we remark that the analogous substitution *cannot* be performed on either of the other two cross-coupled NAND gate pairs. For example, it turns out that it is *not* possible to explain the operation of the circuit in terms of Fig. 6b. This is because *other* properties of this particular NAND gate pair are required, over and above the fact that it behaves as a set-reset flipflop.

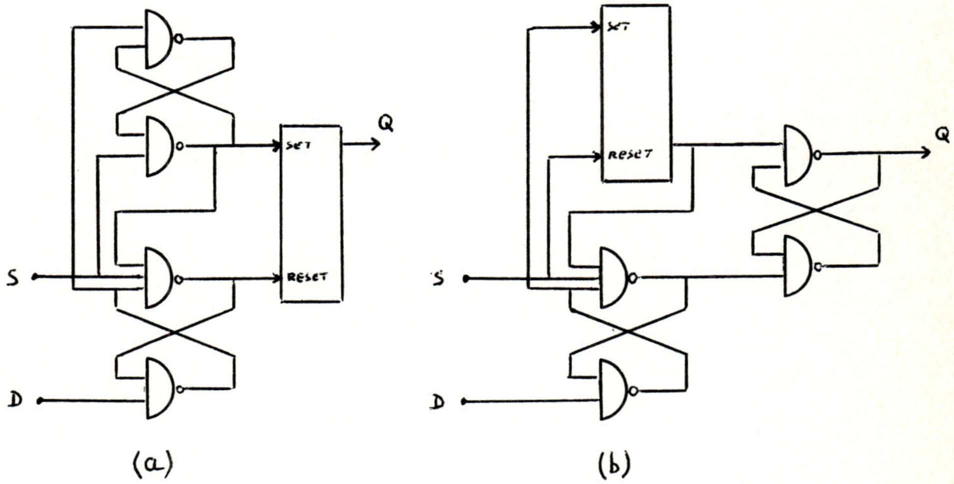

Fig 6.

(a) The circuit shown in Fig 5 split into two subcircuits, the second of which has been replaced by a set-reset flipflop.

(b) An equally plausible, but unsatisfactory substitution.

We will now offer a (very abbreviated) explanation, couched in the informal, everyday jargon of digital engineering, for the operation of the circuit shown in Fig. 6a. We start off

* If, like most people, you find this remark difficult to accept at face value, read the rest of this account, then set it aside, and attempt, within (say) *one working day*, to come up with a carefully justified account of 'how' the proposed implementation is intended to function ...

with the interesting observation that the mode of operation of the circuit is *essentially different* for each of the two polarities of data input. Understanding how the circuit operates when it latches a *high* data input gives little help in the understanding how it operates in latching a *low* data input. The following informal explanation deals only with the former case: we leave the latter case (which turns out to be a little more complex) as an exercise for the interested reader.

4.2 Informal explanation

In order to illustrate the following explanation, we have arbitrarily** chosen certain timing parameters for the NAND gates, and certain timing parameters for the overall flipflop specification. Using these parameters, we have then constructed (see Fig. 7) a set of wavespecs which illustrate (for a *high* data input) the operation of the circuit. In the following, highly informal, account we treat these wavespecs *as if* they were waveforms. The labelling $(a, \ldots, f)$ on the time axis of this diagram corresponds to the following itemisation:

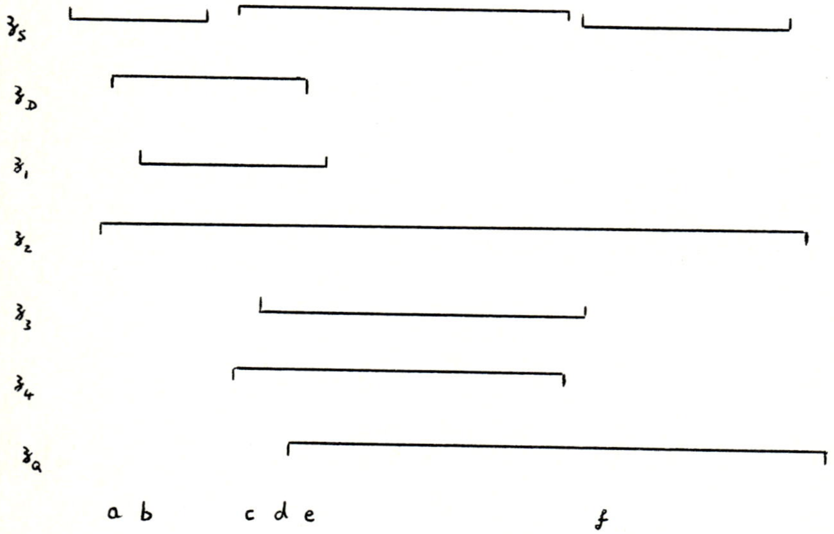

Fig 7. A set of wavespecs compatible with the circuit shown in Fig 5.

(a) Initially, w_S (the strobe waveform) is *low*. Thus w_2 (the output of gate 2) is *high*.

** As it happens (see §7.4) the circuit only functions 'correctly' for certain combinations of timing parameters, so it is *not* an entirely arbitrary choice!

(b) Then, w_D (the data waveform) becomes *high*. Gates g_1/g_2 latch this state. w_1 goes *low*.

(c) Then w_S (the strobe waveform) goes *high*. This drives w_3 *low*. Gates g_3/g_4 latch this state.

(d) w_2 remains *high* and w_3 is *low*. This *sets* the output latch g_5/g_6, with w_Q *high*

(e) w_D (the data waveform) becomes undefined. w_1 also becomes undefined. However, w_3 remains latched *low*, and hence maintains w_2 *high*. Thus output latch g_5/g_6 remains *set*.

(f) w_S (the strobe waveform) goes *low*. This change races through g_2 and g_3: assuming it is a sufficiently rapid transition, w_2 remains *high* and thus the output latch g_5/g_6 remains *set*, with w_Q *high*. (Most of the other gates go into an undefined state).

Again, we emphasise the highly informal, and indeed quite unconvincing, nature of this account. In fact, this example provides an excellent illustration of the *raison d'etre* for formal verification.

4.3 Formal specification of the implementation

Elsewhere ([4]) we have explained how, using higher-order logic, the *structure* (ie, the circuit diagram) of a device may be specified, in the form of an axiomatic theory, and how (by formal inferencing) the corresponding behavioural specification may be obtained from it. Thus, we will not dwell at all upon this aspect in the present account but rather take as our starting point a *behavioural* specification for the circuit.

In order to frame this as a formal *definition* we assume that the waveforms are named $w_S, w_D, w_Q, w_1, w_2, \ldots, w_5$ (as shown in Fig. 5). Further, we assume that the timing parameters of the component NAND gates are described by $\vec{n}$, a 4-tuple of durations. Under these assumptions, we can write down a set of assertions satisfied by these waveforms. Each component NAND gate imposes a relation between various of these waveforms. For example, gate g_1 imposes the relation

$$((NAND\_BEHAV \quad \vec{n}) \quad w_2 \quad w_D \quad w_1)$$

and gate g_2 the relation

$$((NAND3\_BEHAV \quad \vec{n}) \quad w_3 \quad w_S \quad w_1 \quad w_2)$$

The relation imposed by all the gates together is the conjunction of all the component relations.

At this point we introduce a new symbol to denote the wavespec (or, more precisely, the *weakest* wavespec) satisfied by the waveforms. Because the component wavespecs are parameterised by $\vec{n}$, so also this overall wavespec will be. Thus, we introduce the symbol

$$IMPL\_BEHAV \quad : \quad dur^4 \rightarrow (wf^8 \rightarrow bool)$$

(a parameterised wavespec for the behaviour of the implementation) with definition

$$
\begin{aligned}
&((IMPL\_BEHAV \quad \vec{n}) \quad w_S \quad w_D \quad w_Q \quad w_1 \quad w_2 \quad w_3 \quad w_4 \quad w_5) \\
=_{def}& \\
&((NAND\_BEHAV \quad \vec{n}) \quad w_2 \quad w_D \quad w_1) \\
\wedge& \\
&((NAND3\_BEHAV \quad \vec{n}) \quad w_3 \quad w_S \quad w_1 \quad w_2) \\
\wedge& \\
&((NAND\_BEHAV \quad \vec{n}) \quad w_4 \quad w_S \quad w_3) \\
\wedge& \\
&((NAND\_BEHAV \quad \vec{n}) \quad w_1 \quad w_3 \quad w_4) \\
\wedge& \\
&((NAND\_BEHAV \quad \vec{n}) \quad w_Q \quad w_2 \quad w_5) \\
\wedge& \\
&((NAND\_BEHAV \quad \vec{n}) \quad w_3 \quad w_5 \quad w_Q)
\end{aligned}
$$

The waveforms associated with the circuit shown in Fig. 5 (ie, the proposed implementation of a D-flipflop) will therefore (assuming the tuple of timing parameters of each of the component NAND gates is $\vec{n}$) satisfy the wavespec

$$(IMPL\_BEHAV \quad \vec{n}) \quad : \quad wf^8 \rightarrow bool$$

The above definition will be used in the next section

5. Formal verification

We now move onto the second major theme of this paper, that of the 'formal verification' of the correctness of the proposed implementation of the D-flipflop. In principle, the task is a very simple one.

- On the one hand, we have (for an arbitrary 8-tuple $\vec{a}$ of durations) a wavespec

  $$(DFF\_BEHAV \quad \vec{a}) \quad : \quad wf^3 \rightarrow bool$$

  which describes the relation that we *require to be satisfied* between the triple of waveforms w_S, w_D, w_Q at the ports of a D-flipflop.
- On the other hand, we have (for an arbitrary 4-tuple $\vec{n}$ of durations) a wavespec

  $$(IMPL\_BEHAV \quad \vec{n}) \quad : wf^8 \rightarrow bool$$

  which describes the relation that we *know to be satisfied* between the 8-tuple of waveforms $w_S, w_D, w_Q, w_1, \ldots, w_5$ at the ports ('internal' and 'external') of the proposed implementation.

What we need to do, in order to 'verify the design' (that is to prove its correctness w.r.t. the behavioral definition for a D-flipflop) is (roughly stated) to establish that *any* set of waveforms which satisfies the second of these wavespecs also satisfies the first.

Stated a little more carefully, as a formula, this would be:

$$\begin{array}{l}\forall w_S, w_D, w_Q, w_1, w_2, w_3, w_4, w_5 : wf. \\ \quad ((IMPL\_BEHAV \quad \vec{n}) \quad w_S \quad w_D \quad w_Q \quad w_1 \quad w_2 \quad w_3 \quad w_4 \quad w_5) \\ \quad \rightarrow \\ \quad ((DFF\_BEHAV \quad \vec{a}) \quad w_S \quad w_D \quad w_Q)\end{array}$$

Of course, this formula is *not* going to be true for arbitrary values* of $\vec{a}$ (the 8-tuple of timing parameters of the flipflop) and $\vec{n}$ (the 4-tuple of timing parameters of the component NAND gates). Thus, we will postulate a relation

$$R \quad : \quad dur^8 \times dur^4 \rightarrow bool$$

that will define the necessary relationship between $\vec{a}$ and $\vec{n}$ if the above formula is to be true. We shall call R the '*timing relation*'. Using R, we can now state the actual theorem that we wish to establish

$$\begin{array}{l}\forall \vec{a} : dur^8, \ \vec{n} : dur^4. \\ \quad (R \quad \vec{a} \quad \vec{n}) \\ \quad \rightarrow \\ \quad \forall w_S, w_D, w_Q, w_1, w_2, w_3, w_4, w_5 : wf. \\ \qquad ((IMPL\_BEHAV \quad \vec{n}) \quad w_S \quad w_D \quad w_Q \quad w_1 \quad w_2 \quad w_3 \quad w_4 \quad w_5) \\ \qquad \rightarrow \\ \qquad ((DFF\_BEHAV \quad \vec{a}) \quad w_S \quad w_D \quad w_Q)\end{array}$$

We shall call this (as yet unproven) theorem the *D-flipflop implementation theorem*. In words, it states: "**For any** 8-tuple $\vec{a}$ and 4-tuple $\vec{n}$ of durations, **if** they satisfy the condition $(R \quad \vec{a} \quad \vec{n})$, **then any** set of waveforms that occur at the nodes of the proposed implementation (whose NAND gates have timing parameters $\vec{n}$) will meet the specification of a D-flipflop (with timing parameters $\vec{a}$)."

General Aims The example we have just described is entirely typical of hardware verification. Looking at it, we see that we have, in fact, two distinct but interrelated problems to solve. One is evidently to prove (formally or informally, as the circumstances warrant) the above theorem. The other (not so evident) task, however, is to find a definition for the timing relation R. (Notice carefully that we do *not* imply that such a relation is unique, nor do we imply that a non-trivial one exists.) Although one might imagine that proposing a suitable relation is an easy task, experience contradicts this. It turns out that determining a 'suitable' value for R is as difficult a task as proving the above theorem. In our experience, 'intuitive' attempts at deciding a value for R are *invariably* wrong!

---

* For example, it is intuitively obvious that a 'fast' flipflop cannot be realised in terms of 'slow' NAND gates.

The best way that we have found to determine R is to begin by leaving it undefined, and then to attempt to prove the desired theorem. During the course of this proof, one will be forced to make certain 'assumptions' in order to carry through the proof. Then, when the (conditional) proof has been achieved, one can gather up all the assumptions (ie, relations between timing parameters) and *chose* a relation R which is

- sufficiently *strong* to imply all the required assumptions, yet
- sufficiently *simple* to be of practical use (no one is going to be interested in a relation that takes a couple of pages to express!).

The proof can then be reworked with this definition of R present as an axiom.

6. Computational Inferencing

In this section, we review the means by which we carry through the two interrelated aims described above, namely:

- Determining a 'suitable' definition for the timing criterion R, and
- Proving the D-flipflop implementation theorem.

As we shall see, this involves constructing a large and complex proof. We undertake this task 'computationally', in order both to ensure the *correctness* of our inferencing (by eliminating the possibility of 'slips', wishful thinking, etc) and to be able to automate the lower-level aspects of it.

The approach we use originally derives from a method pioneered in the Edinburgh ML/LCF project [1]. In this account, we only have space to give a most cursory overview. The approach is very fully described in [5] (which includes an extensive tutorial exposition).

General Principles A *derivation* (or 'proof') is a *tree*. The *nodes* of the tree are labelled with the primitive *rules of inference* of the logic (for example, *modus ponens*, *generalise*, *specialise*, are some of the rules of inference of the VERITAS logic). The *arcs* of the tree are labelled with *theorems*. The *tips* of the tree are labelled with *axioms*. A derivation is not a free-standing entity, but rather it exists only with respect to a *signature*. A signature (sometimes called a 'theory presentation') is essentially a sequence of *declarations* of the *symbols* and *axioms* that may occur in a derivation.

The VERITAS logic is *computationally implemented* within a programming language MV (for 'META-VERITAS'). This is an interactive, typed, purely functional programming language. The implementation of this language includes (amongst others) abstract data types (ADT's) for *signatures*, *terms* and *derivations*. Associated with these ADT's are functions for constructing, manipulating and displaying values of each type. These functions, which have been very carefully defined and implemented, are such as to *guarantee* the syntactic correctness and the logical soundness of any entity constructed. For example, consider the MV function

$$spec \; : \; deriv \times term \rightarrow deriv$$

which implements the '*specialise*' rule of inference. The function takes two arguments; a derivation and a term. The intention is that the first argument should be a derivation of a theorem of the form

$$\vdash \forall x : s. \quad tm$$

and the second a term tm' of type s. In this event, then the value returned by this function call will be a derivation of a theorem of the form

$$\vdash tm[tm'/x]$$

(ie, the term tm with all free occurrences of the symbol x replaced by the term tm', under the assumption that no capture occurs). Incorporated within the implementation of the ADT function *spec* (and completely out of sight to the user) is a rigorous set of checks (on the types and values of the two arguments) that ensures that, unless the resultant derivation would be correct in every respect, an error condition arises. Similar checks are built into each of the other primitive constructor functions.

6.1 High-level Inferencing

Whilst it is, in principle, possible to construct a derivation of any theorem simply in terms of the primitive ADT term and derivation constructors described above, it would be unimaginably tedious to do so. There are three techniques that may be used to reduce the effort required by a user:

- Functions which implement *derived* inference rules may be written in the meta-language. The body of such a function is a complex expression which, each time an application of the function is evaluated, results in the application of many primitive term and derivation constructors. As an illustration, a very simple example of a derived inference rule would be a function which, given a derivation of a theorem of the form $\vdash \forall x.\ \forall y.\ \forall z.\ P_{x,y,z}$ yields a derivation of the theorem $\vdash \forall x, y, z.\ P_{x,y,z}$, that is, it *tuples* a nested sequence of quantifiers.
- A 'database' of useful lemmas (ie, subsidiary theorems) may be built up. In general, the axioms that are chosen to characterise a theory will be both very simple and few in number (the two factors helping to ensure that the truth of the axioms is beyond reasonable doubt). Because they are simple and few in number, they cannot encompass the many arithmetic, algebraic and digital engineering relations that we would like to be able to take for granted. Thus, it is desirable to build up a database of a few hundred generally useful lemmas. To give a simple example, the following lemma

$$\begin{aligned}&\vdash \forall a, b, c : nat.\\ &\qquad (a \geq b) \wedge (b > c) \rightarrow (a > c)\end{aligned}$$

  is representative of the couple of dozen lemmas concerning arithmetic ordering relations.

- The third technique we mention is that of *goal-directed* inferencing. The idea is a very simple one. It relies upon the fact that if one knows, in advance, the theorem one is endevouring to prove (ie, the *goal*) it is often easy to obtain from it a collection of *subgoals* with the properties (a) that, individually, they are easier to prove than was the original goal, and (b) that, once they themselves have been established as theorems,the original goal may itself (by the use of a derived inference rule) be established as a theorem. The functions that are used for this purpose are known as *tactics*. A typical example of a tactic is one for proving theorems by the use of *induction*. For instance, given a goal of the form

$$\forall n : nat.\ P_n$$

(where P_n is some formula in which n occurs free), the tactic will yield two subgoals namely

$$P_n[0/n] \qquad \text{and} \qquad [P_n] \quad P_n[(succ\ n)/n]$$

and, once these subgoals have been established as theorems, will use an appropriate derived inference rule to infer the original goal.

We describe the system with which the user interacts in constructing a derivation as a *proof assistant*. Basically, it is a program (written in MV) which allows the user to specify the *top-level* goal, to examine any subgoal (and the declarations of symbols and axioms on its associated signature), and to specify the tactic to be used to achieve each of the subgoals.

It is often useful, especially when undertaking new proofs of an unfamiliar kind, to be able to display, in graphical form, the derivation (or completed part thereof). Thus, the proof assistant program provides the means for doing this; as an illustration of this sort of display, Fig 8 shows a typical derivation (in fact, it is a derivation of the NAND-PAIR lemma that will be used in §7.2 below).

This concludes our very brief look at the logic and the techniques used in its computational implementation. For the remainder of this paper, we will limit ourselves to a relatively informal, high-level description of the goal-directed inferencing process we have used to prove the D-flipflop theorem and obtain the timing criterion R.

7. Proving the D-flipflop Implementation Theorem

So far, we have proposed definitions for the behaviour of a NAND gate and for a D-flipflop, we have illustrated one commonly used implementation of a D-flipflop, we have written down the theorem we would like to establish, and we have outlined the approach we use in computational theorem proving. In this section we shall draw all these threads together and describe how we actually went about computationally proving the theorem and determining the timing relation R.

We have already noted that, even at an *intuitive* level, the mode of operation of the proposed implementation is far from obvious. It will scarcely come as a surprise, therefore,

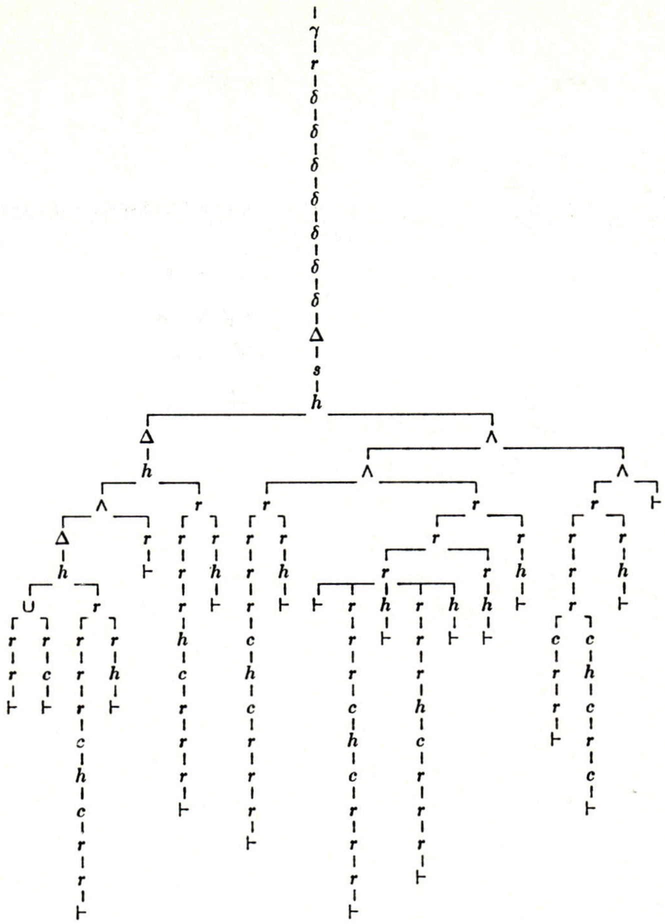

Fig 8. Graphical display of a typical derivation.

that a formal derivation of the D-flipflip implementation theorem is quite complicated. Thus, we are only able to give a general, high-level account of the way we go about constructing this derivation. We mention, however, for anyone who may interested in examining the actual proof, that we are in the process of preparing an annotated step-by-step transcript of it, in the form of a C.A.I interactive document.

Let us start this account of the proof by restating, as our 'goal', the formula we are intending to prove:

$$\forall\ \vec{a} : dur^8,\ \vec{n} : dur^4.$$
$$(R \quad \vec{a} \quad \vec{n})$$
$$\rightarrow$$
$$\forall\ w_S, w_D, w_Q, w_1, w_2, w_3, w_4, w_5 : wf$$
$$((IMPL\_BEHAV \quad \vec{n}) \quad w_S \quad w_D \quad w_Q \quad w_1 \quad w_2 \quad w_3 \quad w_4 \quad w_5)$$
$$\rightarrow$$
$$((DFF\_BEHAV \quad \vec{a}) \quad w_S \quad w_D \quad w_Q)$$

As we have explained such a goal is not a free-standing entity, but rather has associated with it a signature. The signature declares all the symbols and axioms that are, in effect, 'within scope', that is, that may be used in constructing a derivation of the theorem. The 'earlier' parts of this signature introduce symbols and axioms associated with propositional logic, arithmetic, instants of time, durations, intervals and waveforms (as briefly described in §2 above). The 'later' (and thus more problem specific) parts of the signature include declarations for the symbols $NAND\_BEHAV$, DFF_BEHAV and $IMPL\_BEHAV$, and their defining axioms (precisely as we have defined them in §2.3 of this account). The signature also includes a declaration for the timing condition R, but, as yet, no axiom defining it.

7.1 First Part of the Proof

We now give an itemised account of the main stages gone through in constructing (in goal-directed mode) a derivation of the D-flipflop implementation theorem. In presenting the following account, we explain that it is highly abbreviated, and hence gives a somewhat idealised view.

(1) The first step, before even calling the proof assistant, is to construct the signature (or, alternately stated, to define the theory presentation) with respect to which inferencing will take place. This is simply done by 'extending' an existing (standard) signature with the definitions (as given earlier) for the symbols IMP_BEHAV, and DFF_BEHAV and a declaration (but no defining axiom) for the symbol R.

(2) The 'proof assistant' function (in the meta-language) is then called, with the goal (defined w.r.t the above signature) as its argument.

(3) The first couple of steps are concerned with "stripping-off" (by the use of the appropriate tactics) the universally quantified symbols and the hypothesis part of the outer implication. This results in the subgoal

$$\forall\ w_S, w_D, w_Q, w_1, w_2, w_3, w_4, w_5 : wf$$
$$((IMPL\_BEHAV \quad \vec{n}) \quad w_S \quad w_D \quad w_Q \quad w_1 \quad w_2 \quad w_3 \quad w_4 \quad w_5)$$
$$\rightarrow$$
$$((DFF\_BEHAV \quad \vec{a}) \quad w_S \quad w_D \quad w_Q)$$

The signature associated with this subgoal now includes declarations for the symbols $\vec{a}$ and $\vec{n}$ as 'constants', and for $(R \quad \vec{a} \quad \vec{n})$ as an axiom or 'assumption'.

(3) The universally quantified variables $w_S, \ldots, w_5$ are likewise stripped-off, and then the symbol $IMPL\_BEHAV$ is replaced by the RHS of its definition (remember that, in eventually constructing the derivation, exactly the *opposite* of all these moves takes place). This then yields a subgoal of the form

$$\begin{array}{l} \quad ((NAND\_BEHAV \quad \vec{n}) \quad w_2 \quad w_D \quad w_1) \\ \wedge \\ \quad\quad \vdots \\ \wedge \\ \quad ((NAND\_BEHAV \quad \vec{n}) \quad w_3 \quad w_5 \quad w_Q) \\ \rightarrow \\ ((DFF\_BEHAV \quad \vec{a}) \quad w_S \quad w_D \quad w_Q) \end{array}$$

(5) Again, the hypothesis part of this formula is stripped-off, so that the six $NAND\_BEHAV$ clauses appear as axioms in the signature. Then DFF_BEHAV is replaced by its definition. This yields the subgoal

$$\begin{array}{l} \forall\, c : C. \\ \quad \forall\, t : time,\ d, e : dur. \\ \quad\quad (z_S \quad w_S) \wedge ((z_D \quad c) \quad w_D) \rightarrow ((z_Q \quad c) \quad w_Q) \\ \quad\quad WHERE \\ \quad\quad (z_S \quad w) \quad = \quad \ldots \\ \quad\quad\quad \vdots \\ \quad\quad WHERE \\ \quad\quad t_1 \quad = \quad t + a_1 - a_5 \\ \quad\quad t_2 \quad = \quad t + a_1 + a_2 \end{array}$$

(6) Case analysis on c is carried out: this yields two subgoals. One is for c_t and the other for c_f (ie, for *high* and for *low* data inputs respectively). Here, we will consider only the former case. The case analysis is followed by stripping-off the universally quantified variables, replacing the terms $(z_S \quad w_S)$ and $((z_D \quad c_t) \quad w_D)$ by their definitions and then stripping these off (so that they too become axioms). This results in the subgoal

$$\begin{array}{l} ((const \quad ((\Delta \quad a_7 \quad a_8) \quad i) \quad c_t) \quad w_Q) \\ WHERE \quad i \quad = \quad (I \quad t_2 \quad a_3 + d + a_4 + e) \\ WHERE \quad t_2 \quad = \quad t + a_1 + a_2 \end{array}$$

This concludes the first part of the proof strategy: we now have as our goal the above assertion involving the 'output' waveform w_Q, and a whole set of assumptions (present as axioms in the signature of the goal), including wavespecs on the 'strobe' and 'data' waveforms, w_S and w_D.

7.2 Specialised Techniques

Before continuing with the second part of this proof, we pause to describe some techniques that we have introduced to facilitate proving goals of this general kind. These techniques are:

- the notion of 'propagating assertions through gates',
- tactics that allow this type of operation to be carried out in a goal-directed manner,
- tactics for performing 'cutting and stitching operations' on intervals,
- a technique for retrospectively deciding what the timing criterion R should be.

We deal with each of these in turn.

Propagating assertions through gates A technique that we use extensively is, expressed informally, that of propagating assertions through gates. We illustrate the idea with a simple example. Suppose that w_1 and w_2 are the waveforms at the input ports of a NAND gate and that w_3 is the waveform at its output port. Further, suppose that the assertion

$$((const \quad i \quad c_f) \quad w_1)$$

has been established, that is, the w_1 input waveform is known to be in the *low* state during an interval i. Let us (meta-linguistically) name this assertion α_1 (the subscript '1' informally indicating that it relates to waveform w_1). Then, by using the (axiomatic) definition of $NAND\_BEHAV$, we can infer (essentially by *specialisation* followed by *modus ponens*) the assertion

$$((const \quad ((\Delta \quad d_1 \quad d_2) \quad i) \quad c_t) \quad w_3)$$

This assertion (name it α_3) relates to w_3, the output waveform. Intuitively, we can think of assertion α_3 as being obtained by "propagating assertion α_1 through gate g".

Tactics for 'propagating assertions' In order to be able to use the intuitively appealing style of reasoning described above in a *goal-directed* manner, we must be able to express it as a tactic. At first sight, this may seem difficult because goal-directed reasoning inherently proceeds in a 'backwards' direction, whereas the above style of reasoning is inherently 'forwards'. As the following example shows, however, these two conflicting requirements can be reconciled.

Suppose that, for the simple circuit shown in Fig 9, we wish to establish a goal of the form $\alpha_0 \rightarrow \alpha_n$ (where $\alpha_0, \ldots, \alpha_n$ are assertions on waveforms $w_0, \ldots, w_n$ respectively).

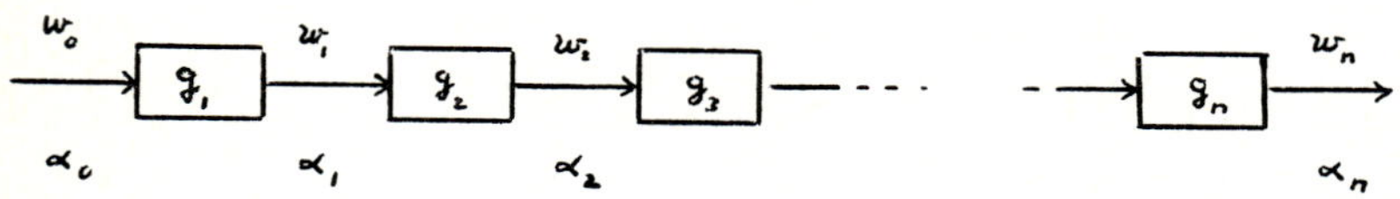

Fig 9 A simple circuit. $\alpha_0, \ldots, \alpha_n$ are assertions involving the waveforms $w_0, \ldots, w_n$ respectively.

The tactic that we use to achieve this operates as follows. First of all, it takes the assertion α_0 and 'propagates' it through gate g_1, thus obtaining a new assertion, α_1 , and a theorem of the form $\vdash \alpha_0 \rightarrow \alpha_1$. The tactic then generates a subgoal of the form $\alpha_1 \rightarrow \alpha_n$. Once this subgoal (which is one step 'easier' to achieve than the original goal) has eventually been achieved (by further subgoaling, etc), it is then a trivial matter for the tactic to infer (essentially by using *modus ponens* and the above theorem) the original goal as a theorem.

This mechanism is, of course, entirely encapsulated within the body of the 'propagate' tactic (or, more precisely, set of such tactics, since there is one for each kind of gate), and so, to a user, the effect is simply as if he was propagating assertions through the circuit in the same manner that a digital engineer is used to thinking of signals as propagating through a circuit.

Tactics for manipulating intervals There is another specialised group of tactics that we also employ extensively: these are used for 'cutting and stitching' the intervals on which wavespecs are defined. For example (see Fig. 10), given a pair of wavespecs z_1 and z_2 (of the same polarity, and assumed to be qualifying the same waveform) one might wish to assume that the intervals on which they are defined overlap, and hence, taken together, imply the wavespec z defined over the union of the two intervals.

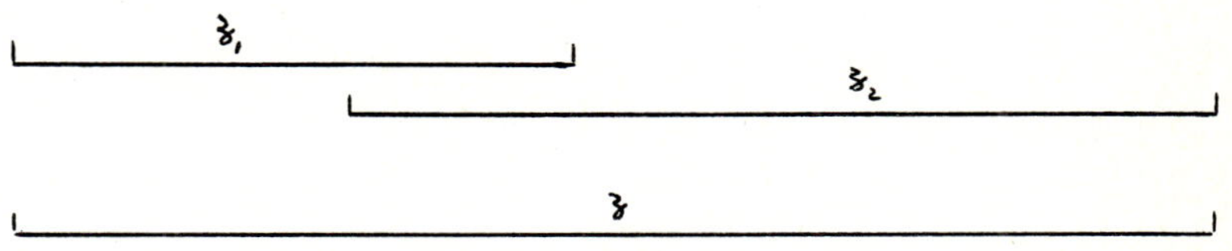

Fig 10 Merging two assertions.

In order to justify this assumption (ie, that the intervals overlap), the proof assistant will generate, as an additional goal, a relation (generally a conjunction of inequalities) that must hold between the two intervals. Because of the number of ways in which intervals may overlap or fail to overlap (including cases where one or other or both intervals are empty), these relations may be unexpectedly complex.

The NAND-PAIR lemma In addition to these specialised tactics, we also make use of a collection of behavioral lemmas. A representative example is one we have named the NAND-PAIR lemma (see Fig. 11). It essentially asserts that a pair of cross-coupled NAND gates behaves like a *set-reset flipflop*, that is to say, it will 'latch' a *low* pulse on one of its inputs, and stay in this state for as long as the other input remains *high*.

In more detail, it asserts that (provided the timing parameters $\vec{b}$ of the flipflop are appropriately related to those of the constituent NAND gates), if the assertions α_{set} and α_{reset} hold, then assertion α_{out} will also hold. In passing, we remark that there are several closely related variants of this lemma.

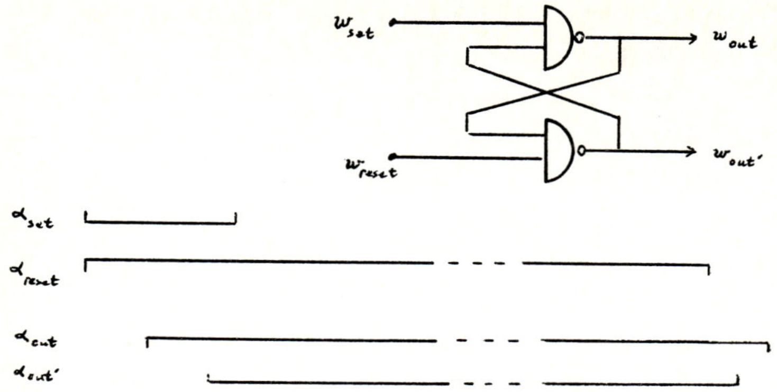

Fig 11 The NAND-PAIR lemma. Assertions α_{set} and α_{reset} imply assertion α_{out}.

Definition of the relation R The net effect of using the various tactics and lemmas we have just been describing is that, as the goal-directed construction of the proof tree proceeds, subsidiary subgoals involving relations between the timing parameters are generated. In general, we do not attempt to satisfy these subgoals immediately, but rather leave them as 'stubs' (ie, as unfinished branches of the goal tree). Then, when all other parts of the goal-tree have been filled in, we gather together the entire set (call it Σ) of such relations and determine a 'suitable' definition for the relation R. Obviously, one possible definition for R would simply be the conjunction of all the relations in Σ, but (because Σ is quite large) such a definition would be far too complex to be of any practical interest. Instead, what we do is to examine Σ carefully and choose a definition for R which is both *reasonably simple* and *sufficiently strong* (ie, strong enough to logically imply each of the relations in R). This takes a degree of skill and judgement. With R chosen, we then incorporate its defining axiom in the signature, and revisit the stubs (ie, unfinished branches) of the goal tree and complete them.

This concludes our summary of the general techniques we use during the second half of the proof. We will now continue the account of the construction of the D-flipflop theorem.

7.3 Second Part of the Proof

The second part of the proof is entirely concerned with 'propagating assertions' through the proposed implementation of the D-flipflop. The overall aim is, briefly stated, to demonstrate that the implementation does, in fact, succeed in 'latching' the input signal. (Recollect that, of the two possible polarities of input signal, only the c_t case is presently under consideration; the c_f case is the subject of a separate proof, not described here.) Stated in a little more detail, the aim is to demonstrate (by goal-directed inferencing) that

under the assumption that the strobe waveform w_S satisfies the strobe wavespec z_S,

and the data-in waveform w_D satisfies the data-in wavespec $(z_D \quad c_t)$, **then** the output waveform w_Q will satisfy the output wavespec $(z_Q \quad c_t)$.

Call these three assertions (or, more precisely, the expanded form of these three assertions, as defined in §7.1 above) α_S, α_D and α_Q respectively. The assertion α_Q is the current goal, namely

$$((const \quad ((\Delta \quad a_7 \quad a_8) \quad i) \quad c_t) \quad w_Q)$$
$$WHERE \quad i \; = \; \ldots$$

and the assertions α_S and α_D are present as axioms (ie, 'hypotheses') in the signature of the goal.

The outline of this part of the proof, which involves the detailed consideration of a series of related wavespecs, is easy to follow if presented in graphical form (see Fig. 12). Essentially, we will be propagating assertions α_S and α_D through the proposed implementation of the D-flipflop (as shown in Fig 5), and demonstrating that they imply assertion α_Q.

In the following description, we use the notation α_i^j to denote the jth assertion relating to waveform w_i.

α_S This assertion specifies the strobe waveform, w_S. It is an *assumption*, that is, it is an axiom present in the signature associated with the current goal.

α_D Similarly, this assertion specifies the data waveform, w_D. Again, it is an assumption.

α_S^1 This assertion is simply the first conjunct of assertion α_S.

α_2^1 This assertion (the *first* one relating to w_2, the waveform on the output of gate $g2$) results from propagating assertion α_S^1 through NAND gate g_2. (*Note:* Less colloquially, we would say that assertion α_2^1 may be inferred from assertion α_S^1 and the axiomatic definition of the wavespec $NAND\_BEHAV$.)

α_2^2 This assertion is simply the assertion α_2^1 restricted to a subinterval whose start has been chosen to coincide with the start of the interval associated with assertion α_D.

α_2^3 This assertion is established using assertions α_D and α_2^2, together with a variant of the NAND-PAIR lemma described above. The two gates which in this case form the pair are g_1 and g_2.

α_1^1 This assertion results from propagating the *two* assertions α_D and α_2^3 through NAND gate g_1. (Compare this with the earlier case where a *single* assertion was propagated through a NAND gate. In the present case, the input assertions are *high* ones, and hence an assertion is needed for *each* of the (two) inputs to the gate. In the earlier case, which involved a *low* assertion, only a single such one was required. This difference is due to the fact that the defining axiom for the wavespec $NAND\_BEHAV$ (see §2.3, earlier) has disjunctive hypotheses for *low* input assertions but conjunctive ones for *high* input assertions.)

α_2^4 This assertion results from propagating the *low* assertion α_1^1 through gate g_2.

α_2^5 This assertion results from 'joining together' assertions α_2^1 and α_2^4. *Note:* We are able to join them together because

(a) The two assertions relate to the same waveform, namely w_2,

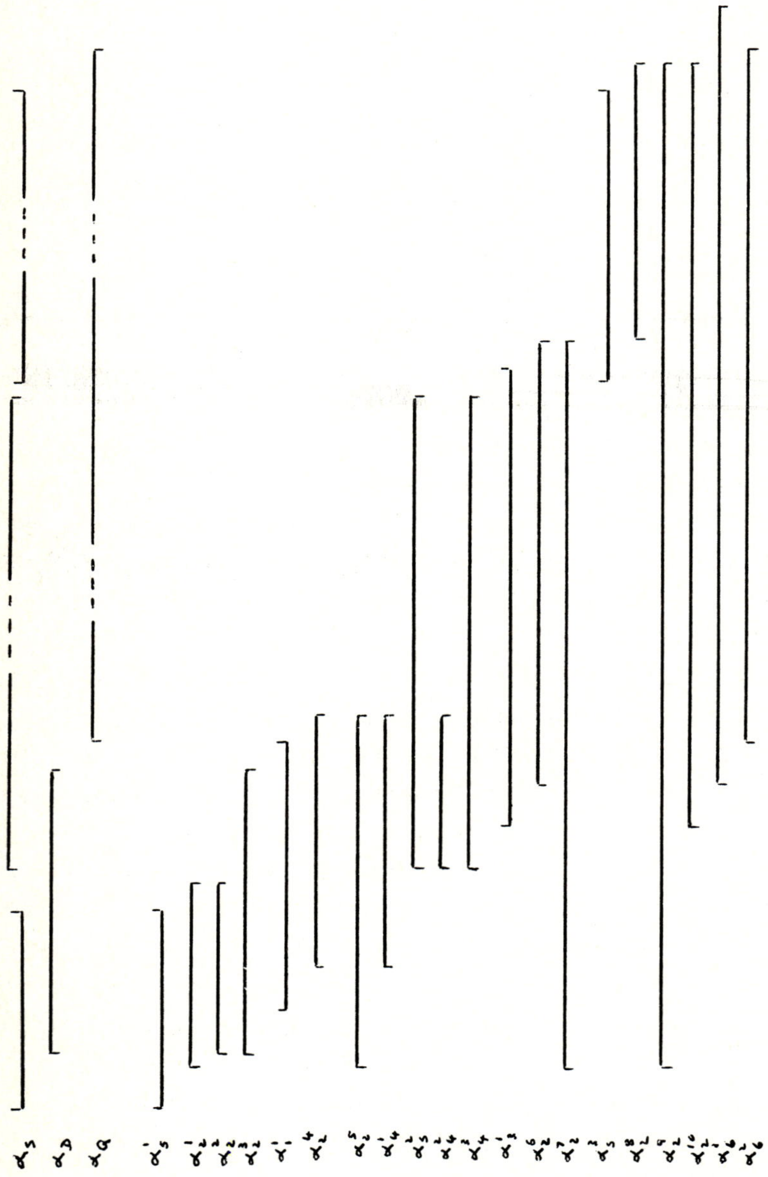

Fig 12 Assertions used in proving the D-flipflop implementation theorem.

(b) Both assertions have the same polarity (in this case, both being *high*), and

(c) The intervals on which the two assertions are defined overlap (or, more precisely, we *require* them to overlap, and, as a result, generate (as further subgoals) constraints on the timing relation, *R*).

α_4^1 This assertion results from propagating assertion α_1^1 through gate g_4.

α_S^2 This assertion is simply the second conjunct of assertion α_S (one of the hypotheses) on the strobe waveform.

α_4^2 This assertion is a weakened version of α_4^1, obtained by restricting the interval over which the latter is defined to a subinterval. This subinterval is chosen so that its start coincides with the start of the interval over which α_S^2 is defined.

α_4^3 This assertion is established using a variant of the NAND-PAIR lemma (this time, with gates g_3, g_4), together with assertions α_S^2 and α_4^2.

α_3^1 This assertion results from propagating the *high* assertions α_S^2 and α_4^3 through gate g_3.

Whilst we could proceed to describe the remainder of the wavespec diagram in these term, it would be tedious to do so. Indeed, the reader may already have perceived that we are using, in a stereotyped fashion, a relatively small number of techniques (ie, tactics). Thus, we can express these operations rather more succinctly. Here (using a notation which is reasonably self-evident) is the remainder of this sequence of assertions.

$$\begin{aligned}
\alpha_2^6 &= \mathit{propagate}\ \ \alpha_3^1\ \ g_3 \\
\alpha_2^7 &= \mathit{join}\ \ \alpha_2^5\ \ \alpha_2^6 \\
\alpha_S^3 &= 3^{rd}\ \mathit{conjunct}\ \ \alpha_S \\
\alpha_2^8 &= \mathit{propagate}\ \ \alpha_S^3\ \ g_2 \\
\alpha_2^9 &= \mathit{join}\ \ \alpha_2^7\ \ \alpha_2^8 \\
\alpha_2^{10} &= \mathit{weaken}\ \ \alpha_2^9 \\
\alpha_6^1 &= \mathit{nand_pair'}\ \ \alpha_3^1\ \ \alpha_2^{10}\ \ g_5\ \ g_6 \\
\alpha_6^2 &= \mathit{weaken}\ \ \alpha_6^1
\end{aligned}$$

This final assertion, α_6^2 is the one we require; it is identical with α_Q, the assertion on the output waveform of the D-flipflop that figured in the original goal.

Thus, (subject only to satisfactorily defining the relation *R*) we have shown that the proposed implementation does indeed function as a D-flipflop for signals of polarity c_t. We remark that the corresponding proof, for signals of polarity c_f, is broadly similar, but a little more complex.

7.4 The Timing Relation

As already described, the end result of undertaking the goal-directed proof construction exercise is not only to prove the D-flipflop theorem, but also to assist in determining a suitable value for R, the timing relation. The actual value that was chosen (and proven correct) for R was

$$
\begin{aligned}
(R \quad \vec{a} \quad \vec{n}) \;=\; & a_1 > 2n_1 + 2n_3 \\
& a_1 \geq a_5 + n_1 \\
& a_3 > 2n_1 + 2n_3 \\
& n_4 + n_2 \geq a_4 + n_1 \\
& a_5 > 2n_1 + n_3 \\
& a_6 > n_1 + n_3 \\
& a_7 \geq n_1 + 2n_3 \\
& n_4 + n_2 \geq a_4 + a_8 \\
& n_2 > 0
\end{aligned}
$$

By way of illustration, we present below a typical 'model' of this relation (ie, a set of values satisfying it). The values we have chosen are representative of a state-of-the-art CMOS gate-array technology.

Timing Parameters for the NAND gate

n_1	=	300 ps	low-to-high propagation delay
n_2	=	200 ps	high-to-low hold time
n_3	=	300 ps	high-to-low propagation delay
n_4	=	200 ps	low-to-high hold time

Timing Parameters for the D-flipflop

a_1	=	1400 ps	min strobe low time
a_2	=	300 ps	max strobe rise time
a_3	=	1300 ps	min strobe high time
a_4	=	100 ps	max strobe fall time
a_5	=	1000 ps	data setup time
a_6	=	700 ps	data hold time
a_7	=	900 ps	max output propagation delay
a_8	=	300 ps	min output hold time

It is of interest to note that

- The 'strobe rise time' parameter, a_2, is (in the definition of R) completely unconstrained: thus, the D-flipflop turns out not really to be 'edge triggered' at all!
- But, conversely, the 'strobe fall time' parameter, a_4, is very strongly constrained; thus, the ability of the D-flipflop to retain data is *critically* dependent upon the rapidity of the trailing edge of the strobe waveform.

So far as we know, this latter fact has not been previously remarked upon.

Functional relationship Although generally there are advantages in describing (as we have done here) the relation between $\vec{a}$ and $\vec{n}$ *relationally*, for some purposes there are advantages in strengthening this relation to a *functional* one. This may take either the form

$$\vec{n} = (F \quad \vec{a}) \qquad \text{or} \qquad \vec{a} = (G \quad \vec{n})$$

The first form specifies the parameters required of the NAND gates in order to realise a D-flipflop with given parameters, whilst the second specifies the parameters of the D-flipflop in terms of those of its component NAND gates (corresponding to, respectively, a top-down *versus* a bottom-up approach). In general, neither F nor G will be total functions.

7.5 Some Statistics

In order to allow the reader to gain an appreciation of the size of the proof of the above theorem, and of the time taken in its creation, we provide a brief summary of the main statistics:

- The overall proof involves about 1600 'high level' inferences (ie, applications of derived inference rules) each corresponding to the selection by the user of a particular tactic.
- These high-level inferences are mapped into about 5000 applications of the 'primitive' inference rules.
- The construction of this proof took about one week of concentrated work by a practiced user.

Note that these figures apply to the proof of the D-flipflop theorem itself; they do not include the proofs of the various lemmas used.

8. Conclusion

The conclusions we draw from the VERITAS project are severalfold; we group them together under various headings. Firstly, as concerns the use of typed, higher-order logic as a computer hardware description language:

- The fact that it allows *predicates* to be introduced means that *partial* (as distinct from *total*) descriptions may be expressed.
- The fact that it allows *higher-order* entities to be introduced means that concepts such as, for example, functional composition or induction, may be expressed in a natural, intuitive style.
- The fact that it has an associated *deductive calculus* means that reasoning ('formal inferencing') can be carried out.
- The fact that higher-order logic is powerful enough to describe mathematics itself means that this reasoning can encompass not just the 'low-level' aspects of the behaviour of digital systems, but also abstract aspects of their behaviour. For example, the properties of number-theoretic transforms based on finite-field arithmetic could be verified.
- Finally, we note that higher-order logic includes as proper subsets both the *applicative programming languages* and also PROLOG. Both of these formalisms have been used for describing aspects of digital systems.

Overall, we would claim that typed, higher-order logic (and syntactic variants thereof) is *the* natural computer hardware description language.

Next, we offer some comments on the nature of low-level digital behavioural specifications.

- These should *always* be partial rather than total. This is because the physical world is inherently indeterminate and our knowledge of it is necessarily imprecise. Any *exact* description must therefore be a wrong one!
- It turns out that the 'temporal' and the 'functional' aspects of low-level behavioural specifications are *inextricably* bound together.
- The predicate *const*, that asserts of an *analogue* waveform that, over a given interval of time, it is constant at a given *digital* state, is of fundamental importance in specifying the behaviour of digital devices.

Our general conclusions on using higher-order logic for formal verification are as follows.

- 'Formal verification', in this context, essentially involves deriving (by the use of formal inferencing, from a set of axioms) a theorem that asserts that the waveforms at the ports of a system satisfy a given property (the behavioural specification of the overall circuit) provided that the waveforms at the ports of each component device of the system satisfy the behavioural specification of that device.
- The process of 'formally verifying' a design is much harder work than is the 'simulation' of the operation of the same device. Unlike simulation, however, the solution obtained from formal verification is valid for *all possible* waveforms and for *all possible* timing parameters.
- Because of this, the theorems expressing the results of formal verification are 'reuseable' items. In particular, this allows the verification of hierarchically structured systems to be undertaken in a structured manner.
- In practice, it is necessary to use *computational assistance* for formal verification in order to eliminate the possibility of wishful thinking, to automate routine patterns of inferencing, and to maintain a secure database of lemmas.
- We have found that an approach to theorem proving derived from the Edinburgh ML/LCF project, in which terms, derivations and signatures are represented as values of secure data abstractions, to be ideal in all respects.

Finally, the conclusions we draw from the particular case study presented here (ie, the specification and verification of a D-flipflop), are as follows:

- When examined in detail, the operation of the proposed implementation of the D-flipflop (which, at first sight appears to be a relatively simple circuit) turns out to be surprisingly complicated.
- The practicability of formally verifying such a circuit has, however, been demonstrated, as also has the practicability (we should rather say the necessity !) of carrying out this task computationally.
- As we have shown, the timing relation R, which relates the timing parameters of the component devices to the timing parameters of the overall circuit, can best be determined by attempting to construct a proof of the verification theorem, gathering together all the 'dangling' subgoals that remain, and then choosing a value of R that is both *strong* enough

to enable the proof to be completed and yet *simple* enough to be of practical utility.

- Finally, in comparing this case study with the many accounts that have appeared in the literature concerning the verification of digital systems, particular note should be taken of the fact that in the approach we have demonstrated:
  - the specifications of both the D-flipflop and its component NAND gates are *partial* ones, and
  - the timing parameters of the component NAND gates have been taken as *variables*, rather than as arbitrary numerical values (eg, unit delays).

  Whilst both these consideration do, of course, greatly complicate the exercise, they are, we believe, the only means by which results of practical use can be obtained.

In overall conclusion, we believe that higher-order logic has been demonstrated to provide an effective basis both for specifying the intended behaviour of digital systems and for formally verifying the correctness of proposed implementations.

Acknowledgment The VERITAS project on digital specification and verification is currently supported under grant GR/C/2287.5 from the UK Science and Engineering Research Council.

9. References

[1] Gordon, M., Milner, R., and Wadsworth, C., "Edinburgh LCF", Springer-Verlag, 1979.

[2] Hanna, F.K., "VERITAS: The Axioms", Internal Report, University of Kent, 1981.

[3] Hanna, F.K., "Overview of the Veritas Project", Internal Report, University of Kent, 1983.

[4] Hanna, F.K., and Daeche, N. "Specification and Verification using Higher-Order Logic", p418, Proc CHDL 1985.

[5] Hanna, F.K., and Daeche, N. "The VERITAS Theorem Prover and its Formal Specification", Internal Report, University of Kent, 1984/5.

[6] Shostak, R.E., "Formal Verification of Circuit Design", p13, Proc CHDL, 1983.

[7] Gordon, M., "Why higher-order logic is a good formalism for specifying and verifying hardware", Internal Report no 77, University of Cambridge, 1985.

[8] Eveking, H., "The Application of CHDL's to the Abstract Specification of Hardware", p167, Proc CHDL 1985.

[9] Herbert, J., "The Application of Formal Specification and Verification to a Hardware Design", p434, Proc CHDL 1985.

[10] Texas Instruments, "Semiconductor Components, Data Book Two",

[11] Enderton, H.B., "A Mathematical Introduction to Logic", Academic Press, 1972.